AF878408

Roland Gyuráki

Fluorescent thermal imaging method for investigating transient effects in high-temperature superconductor tapes and coils

Karlsruher Schriftenreihe zur Supraleitung Band 032

HERAUSGEBER

Prof. Dr. Tabea Arndt
Prof. Dr. rer. nat. Bernhard Holzapfel
Prof. Dr. rer. nat. Sebastian Kempf
Prof. Dr.-Ing. Mathias Noe

Eine Übersicht aller bisher in dieser Schriftenreihe erschienenen Bände finden Sie am Ende des Buches.

Fluorescent thermal imaging method for investigating transient effects in high-temperature superconductor tapes and coils

by
Roland Gyuráki

KIT Scientific Publishing

Karlsruher Institut für Technologie
Institut für Technische Physik

Fluorescent thermal imaging method for investigating transient effects in high-temperature superconductor tapes and coils

Zur Erlangung des akademischen Grades eines Doktor-Ingenieurs von der KIT-Fakultät für Elektrotechnik und Informationstechnik des Karlsruher Instituts für Technologie (KIT) genehmigte Dissertation

von Roland Gyuráki, M.Sc.

Tag der mündlichen Prüfung: 12. November 2019
Referent: Priv.-Doz. Dr. Francesco Grilli
Korreferenten: Prof. Dr. Frédéric Sirois
Prof. Dr.-Ing. Mathias Noe

Impressum

KIT Scientific Publishing

Karlsruher Institut für Technologie (KIT)
KIT Scientific Publishing
Straße am Forum 2
D-76131 Karlsruhe

KIT Scientific Publishing is a registered trademark of Karlsruhe Institute of Technology.
Reprint using the book cover is not allowed.

www.ksp.kit.edu

This document – excluding parts marked otherwise, the cover, pictures and graphs – is licensed under a Creative Commons Attribution-Share Alike 4.0 International License (CC BY-SA 4.0): https://creativecommons.org/licenses/by-sa/4.0/deed.en

The cover page is licensed under a Creative Commons Attribution-No Derivatives 4.0 International License (CC BY-ND 4.0): https://creativecommons.org/licenses/by-nd/4.0/deed.en

Print on Demand 2022 – Gedruckt auf FSC-zertifiziertem Papier

ISSN 1869-1765
ISBN 978-3-7315-1064-2
DOI 10.5445/KSP/1000125489

Acknowledgements

I wish to extend a special thanks to my thesis supervisor Dr. Francesco Grilli for his continuous support, advices and invaluable discussions throughout my PhD. I would also like to thank prof. Mathias Noe for his advices and for suggesting focus points for my research, regarding the needs of the industry and the broader research community. I also thank prof. Frédéric Sirois for his support and comments regarding the experimental setup and measurement procedures using fast, pulsed currents.

I am grateful for the help and support of my colleagues and fellow PhD students at the Institute for Technical Physics, KIT. Dr. Anna Kario, Dr. Sebastian Hellmann, Aurélien Godfrin, Jean-Hughes Fournier Lupien, Fabian Schreiner, Ruslan Popov, Shahab Karrari, Dr. Tara Benkel and Nicolo Riva all helped my research work in the laboratories and with valuable discussions about experimental procedures and the interpretation of results. I would like to also thank Andrea Kling, Andrej Kudymow, Johann Willms, Uwe Walschburger, Uwe Mirasch and Holger Fillinger for their help and support in the laboratory and with setting up the experimental devices and measurements.

I would further like to thank Nicolo Baroni and Marius Jakoby from the Institute of Microstructure Technology, KIT, for their help with fluorescence spectroscopy measurements. Their support and providing access to the available laboratories and facilities made in many aspects this work possible.

Finally, I would like to express my gratitude to my family and friends for their encouragement and unconditional support throughout my thesis.

Kurzfassung

Hochtemperatur-Supraleiter (HTS) der zweiten Generation versprechen sowohl einen höheren Wirkungsgrad als auch eine geringere Masse in elektrischen Anwendungen. Eine der verbleibenden Herausforderungen, die eine Kommerzialisierung dieser Technologien beeinflusst, ist die Stabilität und Betriebssicherheit der Betriebsmittel. Es ist von Niedertemperatur-Supraleitern (LTS) bekannt, dass Störungen einen lokalen Übergang in den normalleitenden, verlustbehafteten Zustand verursachen. Dieser Vorgang wird als Quench bezeichnet. Aufgrund der hier auftretenden hohen Stromdichten kann dies schwerwiegende Folgen mit sich ziehen, bis hin zur kompletten Zerstörung der Betriebsmittel. Obwohl HTS grundsätzlich eine höhere Stabilität als LTS aufweisen, ist gleichzeitig die Quenchausbreitungs-Geschwindigkeit (NZPV) um mehrere Größenordnungen kleiner. Dadurch können "hot spots" entstehen die letztendlich auch die supraleitenden Systeme beschädigen.

Um bessere Schutzmaßnahmen zu entwickeln, die die Stabilität von HTS Bandleitern und Betriebsmitteln positiv beeinflussen, ist ein besseres Verständnis des Quenchvorgangs entscheidend. Wärmebildaufnahmen dieses transienten Quenchvorgangs könnten einen tieferen Einblick in die Mechanismen eines Quenchprozesses und den damit verbundenen thermischen Runaway liefern. Eine solche Messtechnik existiert für die geforderten kryogenen Temperaturen und Zeitauflösungen nicht, könnte sich jedoch als eine neu einsetzbare Methode für die Supraleiter-Gemeinschaft als sehr wertvoll erweisen.

Im Mittelpunkt dieser Arbeit stand die Entwicklung und Demonstration einer Wärmebildmethode, basierend auf fluoreszierenden mikro-thermographischen Aufnahmen, um diese für die Untersuchung von HTS Bandleitern und Spulen im transienten Zustand einzusetzen. Durch die Hochskalierung der fluoreszierenden Wärmebildtechnik, wird hier ein neues Messverfahren demonstriert. Die neue Methode kann Temperaturen von Objekten, außerhalb des mikroskopischen Bereichs, in der 2D-Ebene erfassen. Weiterhin sind die Aufnahmen dank einer kommerziellen Hochgeschwindigkeitskamera mit einer deutlich besseren Zeitauflösung möglich und dadurch erlaubt das Verfahren auch die Erfassung von transienten, thermischen Effekten. Die hier präsentierte hochgeschwindigkeits-fluoreszierende Wärmebildtechnik benutzt europium tris[3-(trifluoromethylhydroxymethylene)-(+)-camphorate] (EuTFC) als fluoreszierenden

Farbstoff mit präzise gemessenen temperaturabhängigen, fluoreszenten Lichtemissionen. Der Farbstoff wurde durch Tröpfchenbeschichtung auf die Oberfläche von Proben aufgebracht, und anschließend einer Wärmebehandlung von 30 min bei 175 °C unterzogen, um die Beschichtung zu stabilisieren. Die beschichteten Proben wurden bis 77 K gekühlt und die fluoreszierende Oberflächenbeschichtung mit UV LEDs angeregt. Im Bereich von 77 K bis 260 K verringert sich die fluoreszente Lichtintensität nahezu linear, wobei die Farbe unverändert bleibt. Um Temperaturen von den Wärmebildaufnahmen berechnen zu können, wurden zuerst sowohl die Kalibrierung des Farbstoffs als auch die Aufnahmen mit der Hochgeschwindigkeitskamera bei einer definierten Temperatur (hier 77 K, Siedepunkt von flüssigem Stickstoff bei Atmosphärendruck) normalisiert. In der Nachbearbeitung lässt sich so die Oberflächentemperatur in den Aufnahmen anhand der Kalibrierkurve bestimmen.

Zur Demonstration wurde das Verfahren zur qualitativen Quenchmessung an HTS Bandleitern sowie zur Bestimmung ihrer NZPV in flüssigem Stickstoff eingesetzt. Hier wurden kurze Stromimpulse, nahe der kritischen Stromstärke, auf mehrere Bänder mit unterschiedlichen Architekturen angewendet, um die auftretenden thermischen Effekte zu visualisieren. Bei Stromimpulsen zwischen 30 % und 50 % über dem kritischen Strom des Supraleiters war eine Erwärmung der Bandleiter in Form von Streifen, gleichzeitig an mehreren Stellen, sichtbar. Dies zeigt, dass die Quenchausbreitung komplexer als eine einzelne normalleitende Zone ist, die sich mit einer gleichbleibenden Geschwindigkeit in eine Richtung ausbreitet. Gleichzeitig kann die Amplitude des Strompulses nicht nur die NZPV, sondern auch das Verhalten des Quench signifikant beeinflussen. Es wurde auch gezeigt, dass Ströme deutlich über dem kritischen Strom in Bandleiter mit Kupferstabilisierung quasi sofort in die Stabilisierung übertragen werden. Stromimpulse rund um den kritischen Strom – sowohl in Leitern ohne Stabilisierung als auch in stabilisierten HTS Bandleitern – verursachen jedoch eine einzige Quenchstelle, die sich ausbreitet und den Messbereich der Wärmebildaufnahmen schnell überschreitet.

Die durch die Wärmebildaufnahmen ausgerechnete NZPV Werte wurden mit elektrischen Messungen von Polytechnique Montréal, Kanada verglichen, um das Verfahren zu evaluieren. Während die beiden Methoden Quench und NZPV aus verschiedenen physikalischen Ansätzen definieren, zeigten die Ergebnisse, dass bei hohen Stromamplituden die NZPV ähnlich waren. Da die Erwärmung mit der zweiten Potenz des Stromes skaliert, wird angenommen, dass in Bandleitern mit hohen kritischen Stromstärken das elektrische Feld und die Erwärmung sich nahezu zeitgleich ausbreiten. In HTS Bändern mit niedrigeren kritischen Stromstärken lieferte das fluoreszente Wärmebildverfahren, durch die deutlich geringere Heizleistung, jedoch konstant niedrigere NZPV Werte.

Die Wärmebildmethode wurde über HTS Bandleiter hinaus, für das Testen von zwei "Pancake Spulen" angewendet. Eine der Spulen wurde als konventionelle isolierte Spule gewickelt, wobei für die zweite Spule die neue nicht-isolierte Wickelmethode benutzt wurde. In den Messungen wurde die Oberfläche der Spulen mit dem fluoreszierenden Farbstoff beschichtet und durch Leitungskühlung im Messaufbau bei 77 K gemessen.

Die Selbstschutzmechanik der nicht-isolierten Spule wurde durch Betrieb im überkritischen Strombereich erfasst. Hier haben die Wärmebildaufnahmen eine schnelle Umverteilung der Ströme (und Heizleistung) innerhalb der Spule gezeigt. In einer anderen Messung wurde ebenfalls ein thermischer Runaway bei 110 % des kritischen Stromes beobachtet. Über einen Zeitraum von ~60 s verursachte ein langsamer aber stetiger Spannungsanstieg einen Stromfluss in die radiale Richtung. Jedoch wurden keine lokalen "hot spots" oder eine ausgeprägte Aufheizung erkannt. Ein schneller Temperaturanstieg zeichnete sich während des Übergangs in den Normalleitenden Zustand ab. Hier haben sich die inneren Windungen der Spule bis zu 100 K erwärmt, was somit den schwächsten Punkt der Spule anzeigt. Die Spule wurde manuell entladen um Schaden vorzubeugen und die Erwärmung konzentrierte sich in der Mitte der Spule und umfasste nur einige der innersten Windungen. Die Ausbreitung der Wärmezone in die Querrichtung war vernachlässigbar.

Keine Erwärmung wurde erkannt in der isolierten Spule während Normalbetrieb. Um einen Schwachpunkt künstlich zu erzeugen und die Stabilität der Spule zu untersuchen, wurde ein elektrischer Widerstand auf die Spulenoberfläche geklebt und als Heizelement benutzt. Nach einem 7 s langen Heizimpuls mit einer Leistung von 3.2 W wurde ein thermischer Runaway erzeugt, bei dem sich die Spule nach Ende des Heizpulses nicht mehr stabilisieren konnte. Der Temperaturanstieg wurde um das Heizelement herum lokalisiert, wo er sich entlang der Wickelrichtung in Kontakt mit dem Heizelement ausbreitete, jedoch nicht in Querrichtung. Es wurden Temperaturen von bis zu 150 K erreicht, woraufhin die Spule schnell entladen wurde. Sowohl die isolierte als auch die nicht-isolierte Spule haben ihren ursprünglichen kritischen Strom beibehalten und zeigten keine Anzeichen von Schäden auf.

Abstract

Second generation high-temperature superconductors (HTS) are an enabling technology for high field magnets and also a promise for higher efficiency and reduced mass in electrical applications. A technological challenge that remains and limits in many cases the more widespread application of HTS is the stability and protection of these devices. It is known from low-temperature superconductors (LTS) that disturbances can cause a local loss of superconductivity and a quick reversal into a normal-conducting, dissipative state. This transition is referred to as quench. Due to large engineering current densities a quench can have severe consequences including the total destruction of the superconducting devices. While HTS are inherently more stable than LTS, their normal zone propagation velocity (NZPV) is also inherently orders of magnitude slower. This means that localized hot spots can still form, which may lead to damaging the superconducting apparatus.

A deeper understanding of the quench is crucial for better understanding the stability of both HTS tapes and devices as well as for designing protection systems. Thermal imaging of this transient effect is expected to give a deeper insight into the mechanism of quench as well as thermal runaway in HTS applications. Such a measurement technique does not exist for the required cryogenic temperatures and time resolution, however it would prove most valuable as a novel measurement technique for the superconducting community.

This work focused on developing a thermal imaging method, based on fluorescent microthermographic imaging, to be used for investigation of HTS tapes and coils in transient states. By up-scaling the fluorescent thermal imaging method to work with objects outside of the microscopic domain as well as implementing it at vastly higher speeds using a commercial high-speed camera, a new measurement method is presented for 2D temperature mapping of superconducting applications at cryogenic temperatures. The introduced high-speed fluorescent thermal imaging method uses europium tris[3-(trifluoromethylhydroxymethylene)-(+)-camphorate] (EuTFC) as the fluorescent dye with a precisely measured temperature dependent fluorescent light emission. The dye is applied as a coating on a surface to be measured using droplet deposition, followed by a 30 min heat treatment at 175 °C for stabilization. After cooling the sample to 77 K, UV LEDs were used for excitation, causing the coating to emit visible light as a function of its temperature. Over a temperature range of 77 K to 260 K the fluorescent light intensity decays linearly, while the colour of the emitted light remains unchanged. By normalizing both a thermal imaging measurement and the temperature calibration of the fluorescent dye around a known reference temperature (for example 77 K, the boiling temperature of liquid nitrogen at atmospheric pressure) in the post-processing, the temperatures can be calculated using the initial calibration curve within this range. The result is a sequence of images showing a surface temperature map of the imaged object.

For demonstration the method was used for qualitative quench measurement in HTS coated conductor tapes as well as to determine their normal zone propagation velocity in liquid nitrogen. Short current pulses around the critical current were applied to several tapes with different architectures and showed effects previously also presented in the literature, however from a thermal perspective. The forming of "banding quench" was visible on several occasions with short current pulses 30 % to 50 % above the critical current of the tapes. This indicates that the NZPV may be more complex than a single zone propagating at a given speed and that the current amplitude drastically changes the behaviour of the quench and not only the NZPV itself. It was also shown that in tapes with a metallic stabilization layer currents significantly above the critical current transfer quasi instantaneously into the shunt layer and cause a uniform heating and film boiling over the surface. Current pulses around the critical current in both stabilizer-free and stabilized tapes caused a local, propagating zone, that quickly exceeded the temperature limit of the thermal imaging.

The results of NZPV calculated from the thermal imaging were compared with electrical measurement data provided by Polytechnique Montréal, Canada for validation. While the two methods define quench from different physical approaches, the results indicated that at high current amplitudes the NZPV were similar. It is hypothesised that since the heating scales with the second power of the current, in tapes with a high critical current the electric field and heating develop closely simultaneously. In tapes with a lower critical current, however, the thermal imaging reported consistently slower propagation speeds due to the significantly lower heating power.

The method was also implemented beyond single HTS tapes on two demonstrator pancake coils, wound with and without turn-to-turn insulation. In these experiments the coils' finished surface was coated with the fluorescent dye and were measured in a conduction-cooled assembly at 77 K. In the non-insulated coil the self-protecting behaviour was observed in an experiment where a sudden change in current paths was recorded together with a quick redistribution of the heating on the thermal imaging. In an overcurrent measurement at 110 % of the critical current a thermal runaway was also captured. Over a duration of ~60 s a slow yet steady voltage rise indicated current transfer into the radial path, however no local hot spots or distinct heating were identified. A rapid temperature rise became apparent only at the point of the superconducting transition, where the coil's centre windings heated to approximately 100 K, indicating this as the weakest point. While the coil was manually protected to avoid damage, the heating was concentrated around the coil's centre and encompassed merely the innermost few windings. Propagation in the radial path was minimal.

In the insulated coil no heating was detected during normal operation. To create a localized disturbance and determine the stability, a resistive surface heater was retrofitted on the upper side of the coil. A thermal runaway was detected in the coil after a 7 s long 3.2 W heater pulse, where the coil did not stabilize after the heater was turned off. The temperature rise was localized around the heater where it was spreading along the windings in contact with the heater, but not in the transverse direction. Temperatures reached up to 150 K at which point the coil was rapidly discharged. Both the insulated and non-insulated coils have retained their critical currents after the measurements and have shown no signs of damage.

Contents

1 Introduction and motivation

Decades of development and engineering ingenuity led to an efficiency in the range of 85 % to 95 % in today's state-of-the-art electrical motors and generators. In high-performance power transformers efficiencies of up to 99.85 % were also reported [1]. Simultaneously, the average annual losses in the U.S. transmission and distribution grid were estimated at 5 % by the the U.S. Energy Information Administration (EIA) [2]. These extremely high efficiencies are already difficult to challenge, however, there is still scope for improvement for niche applications where for example mass and volume reductions are highly sought-after. There are also technological challenges, which are hard – if not impossible – to overcome with today's conventional electrical technologies.

As wind turbines get bigger both in size and electrical output, direct drive generators are often used to avoid the need of complex and maintenance-hungry gearbox systems. However, with increasing power output the size and correspondingly the mass of the generator increases as well. Heavier generators need a more solid mast and foundation for the wind turbine, which incurs in additional costs especially in the case of offshore wind turbines [3]. High-voltage transformers use SF_6 gas as well as oil for electrical insulation. The former of these is a hazardous greenhouse gas while the latter poses a fire hazard in case of malfunction. On a power transmission level, grid operators are faced with a challenge when disconnecting transmission lines in case of a short-circuit due to high currents.

The use of superconductors, in particular high-temperature superconductors, could be the answer to a range of technological challenges in the electrical sector. Due to their peculiar property of practically zero electrical resistance below a certain temperature, they can replace traditional copper and aluminium conductors. They are promising for electrical applications, such as motors, generators, cables and high field magnets [4–8], due to their high critical current density in high applied magnetic fields and to their mechanical properties. The benefits when compared to conventional machinery and counterparts are increased current density, the possibility of reduction in mass and volume as well as improved efficiency and loss reduction.

Using superconductors for wind turbine generators would result in higher current densities, an increased air gap flux density and consequently reduced size and mass. This can then result in real cost reduction for the foundation and the tower of the wind turbine as well as allow lifting the nacelle in one piece and avoid expensive assembly on top of the wind turbine in harsh environments.

In a superconducting transformer, fabricated from HTS, the dangerous materials could be replaced by a cryogenic liquid that poses no environmental or fire hazard. At the same time, reduction in size and mass can benefit industries such as shipping.

Superconductors are not only able to replace traditional electrical equipment, but also serve as an enabling technology for devices otherwise impossible. Due to the extremely high current densities of superconductors, magnetic fields several times higher than conventional copper coils may be generated. This property was already commercially capitalized upon in several fields. Magnetic resonance imaging (MRI) in hospitals would not be possible at the available resolution without the use of superconductors. The state-of-the-art technology uses low-temperature superconductors (LTS) that can generate strong and stable magnetic fields of typically 1.5 T [9]. The dipole magnets at the European Organization for Nuclear Research (CERN), the magnets responsible for bending the particle beam along the 27 km long tunnel, serve as another example. These generate a magnetic field of 8.3 T and carry a current of over 11 kA [10]. Achieving the same result with traditional conductors would have required an underground tunnel system over four times longer. Besides LTS, HTS applications are also being realized, despite the relative infancy of the technology when compared to LTS. A big advantage – compared to LTS – is that in most cases cooling by liquid nitrogen is possible, at a fraction of the cost compared to liquid helium. In the framework of the AmpaCity project, a 1 km long HTS cable was commissioned in the city of Essen, Germany in 2012 with a transmission capacity of 40 MW [11]. The cable is part of the local grid, connecting two substations and has been operating since installation. Superconducting fault current limiters (SFCL) help solve the problem of fault currents in the grid as well, by acting as a "transparent" component during normal operation. Using the intrinsic physical property of superconductivity, were a sudden increase in line current to happen, the sharp transition between the superconducting and normal states would limit the current to manageable levels and allow for safely disconnecting the affected part of the grid.

Despite the achievements of superconductors many challenges remain. Due to defects in conductors, be it due to the manufacturing process or any other source, cables as well as applications are prone to permanent damage due to the so-called "quench" [12, 13]. A quench is the abrupt transition of a superconductor from truly zero resistance state back into the normal conducting state. Due to high current densities in a superconductor during operation, when a small section of the conductor reverts to normal conductor and poses a finite resistance to electrical current, local heating appears. This local heating, combined with low heat capacities at cryogenic temperatures, can result in a local burnout. The joint effect of heat conduction along the conductor and the continued joule heating cause the local hot spot to propagate at a certain speed, referred to as the normal zone propagation velocity (NZPV). While this speed can be in the range of $\mathrm{m\,s^{-1}}$ in LTS [14], in the case of HTS, it is generally in the order of only a few $\mathrm{cm\,s^{-1}}$ [15, 16], with new techniques being worked on to achieve speeds in the $\mathrm{m\,s^{-1}}$ magnitude as well [17, 18]. Since a slow NZPV is generally correlated to a higher chance of local burnout, quench and quench protection are crucial research focus points at present time. This problem is directly related to magnets as well, where the stored energy is converted to heat during a quench. Better understanding how, where and why a quench happens in a superconducting device is the foundation for a safer and more reliable operation.

For determining the safe and stable operation margins for superconducting devices it is crucial to further investigate the behaviour of HTS coated conductors and superconducting devices during quench and the corresponding NZPVs. Several electrical measurement methods exist for measuring magnetic field and voltage development in applications with high precision. However, so far no method was available for measuring the heating over a 2D surface at cryogenic temperatures that would allow observing the thermal behaviour of superconducting applications in operation. Such a measurement method could answer important questions for the magnet community about the stability and performance of prototype magnets and also prove useful for determining the thermal stability of different appliances.

This work focused on developing a thermal imaging method – specifically for use in cryogenic environments – to map surface temperatures in superconducting tapes and devices at high speeds and over larger surface dimensions. The capability of mapping the temperature distribution over a 2D surface at thousands of frames per second in a cryogenic environment constitutes the main novelty of the method compared to previous works. The developed method consequently allows capturing thermal images of transient effects over surfaces several centimetres in size. To validate and show the merits of the method, HTS tapes were investigated in a liquid nitrogen bath for quench formation and propagation. Apart from demonstrating the method at high speeds the main goal was to provide new insights into the development of the quench itself and an alternative to NZPV measurements done by conventional electrical methods. To further demonstrate the applicability of the presented thermal imaging technique, the experiments were extended to two superconducting "pancake" coils; a conventional coil with electrical insulation between the turns as well as one wound with a recently introduced technique where the turns of the coil are not insulated from one another. Measuring temperatures in coils at several positions simultaneously is challenging with conventional approaches and hence the aim was to correlate the electrical, magnetic and thermal recordings to get a deeper insight into the coils' thermal behaviour, stability and eventually the normal transition. The thesis is structured as follows.

Chapter 2 gives a brief introduction to superconductivity, focusing mainly on the second generation of HTS tapes. The stability of such tapes is discussed together with stability and particularities of HTS pancake coils, with focus on the no-insulation winding technique presented for the first time in 2011.

Chapter 3 introduces the fluorescent thermal imaging method, applicable at high speeds. The working principles of the fluorescent technique are explained in detail including the preparation, application and calibration of the temperature-sensitive fluorescent coating, the choice of excitation UV light source, data acquisition using a commercial high-speed camera as well as the post-processing method.

In Chapter 4 the introduced thermal imaging technique is implemented to investigate quench behaviour of various HTS coated conductors, in liquid nitrogen, from two perspectives: normal zone propagation velocity and qualitative quench propagation. In the former, thermally calculated normal zone propagation velocities were measured with the new method and compared with those of electrical measurements reported from Polytechnique Montréal, Canada. The second part of the chapter then details particularities about various behaviours of quench – from a thermal perspective – in different HTS tape architectures.

Chapter 5 extends the presentation of the fluorescent thermal imaging beyond single, flat HTS tapes and shows the merits of the method on the example of two HTS-wound pancake coils. One of the coils uses the conventional insulated coil winding technique whereas the second coil was wound without turn-to-turn insulation to investigate the thermal behaviour of this newly introduced, passive quench protection technique. Both coils were tested rigorously in a conduction-cooled assembly while electrical readings as well as thermal images were taken simultaneously and evaluated in detail. The investigation focused on the heating effects due to varying coil current ramping speeds, natural local defects in the conductor, sudden discharge, ac current ripples and finally overcurrent and externally induced heat pulses in the non-insulated and insulated coils, respectively.

Chapter 6 summarizes the findings of the work including the applicability of the introduced fluorescent thermal imaging method and relevant, new findings about quench behaviour of both HTS tapes and coils. An outlook is also presented for the further development of the measurement method highlighting applications of particular interest.

2 Overview of HTS tapes and their stability

This research work focused on developing and implementing a novel measurement technique that can benefit the further development of superconductors and appliances made from superconductors. The detailed explanation of superconductivity, the underlying physics as well as the various types of superconductors is beyond the scope of this work. This chapter merely gives a general introduction into the phenomenon of superconductivity and the concepts related to it that are crucial for understanding the scope of this research. For more in-depth explanation of superconductivity refer to sources such as [7, 19–22].

2.1 Superconductivity

2.1.1 Zero resistance in superconductors

Superconductivity is a phenomenon where a material completely and suddenly loses its electrical resistance below a certain temperature. The discovery was made in 1911 by a Dutch physicist, Heike Kamerlingh Onnes, when he noticed that if mercury was cooled to 4.2 K its resistance could no longer be measured [23].

To better understand the peculiar properties of superconductors, a comparison can be made between the electrical behaviour of common conductors (such as metals), a hypothesised "perfect conductor" (a perfectly pure metal) and a superconductor, at cryogenic temperatures. This is illustrated in Figure 2.1. When a conventional electrical conductor is cooled, its resistance R decreases initially linearly and at a certain temperature reaches a minimum, where further cooling no longer reduces the conductor's resistance. Assuming a perfectly pure metal the electrical resistance would behave in a similar way, however, due to the lack of impurities it might reach zero. Contrary to this, in a superconductor a sudden transition exists at a specific temperature where the resistance abruptly drops to zero.

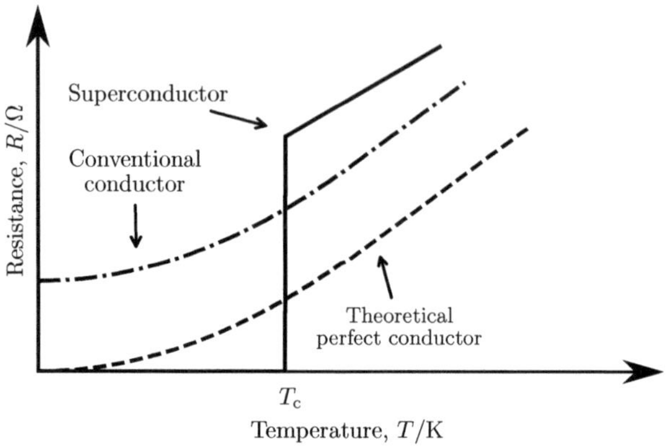

Figure 2.1: Typical behaviour of electrical resistance in conventional conductors, a hypothetical "perfect conductor" and superconductors when cooled to cryogenic temperatures.

2.1.2 Critical parameters

The temperature point where the sudden loss of electrical resistance happens is referred to as the critical temperature T_c of the superconductor as marked in Figure 2.1. However, keeping a superconductor below its critical temperature alone is not sufficient to maintain the superconducting state. There also exists a maximum current density that can be applied to a superconductor without triggering transition into a normal conducting state. This is referred to as the critical current density J_c or often given more conveniently for a given conductor architecture and cross-section as the critical current

$$I_c = J_c A, \tag{2.1}$$

where A is the cross-sectional area of the conductor.

Superconductors are also influenced by magnetic fields, where their critical current density decreases with increasing field strengths. As such, a maximum magnetic field can be defined as well, above which the superconducting state perishes. This is referred to as critical magnetic field H_c, however in the superconductivity community often the critical magnetic flux density B_c is used instead as

$$\mathbf{B} = \mu \mathbf{H}, \tag{2.2}$$

where μ is the permeability of the material. This custom will be adhered to in this work as well and the magnetic field will be reported hereafter in tesla (T).

Considering these three governing parameters: T_c, J_c and B_c and their influence on the critical current, there exists a "critical surface" that can be illustrated as in Figure 2.2. The dark region indicates where the superconducting state persists in a conductor and where crossing the critical surface causes the superconductor to revert into normal conducting state.

2.1.3 Types of superconductors

Since the initial discovery, many pure elements of the periodic table as well as compounds were found to show superconducting properties. Figure 2.3 [24] shows the discovered superconductors from 1911 to 2015 with the name of the compound as well as their corresponding critical temperature.

Superconductors can be grouped together based on several different aspects, however, perhaps the most important distinction can be made based on their critical temperature. Here a separation can be made between low-temperature superconductors (LTS) and high-temperature superconductors (HTS). This distinction has historical reasons, as at the time the prevailing Bardeen–Cooper–Schrieffer (BCS) theory [25] predicted that superconductivity cannot exist above ~30 K. When eventually superconductivity above this limit was discovered in 1986 [26], the new group of superconductors were called HTS.

Most pure element superconductors are called type I, whereas compounds – such as HTS – are usually referred to as type II. This grouping is purely based on the physics of superconductivity and describes the behaviour of the conductor in a magnetic field.

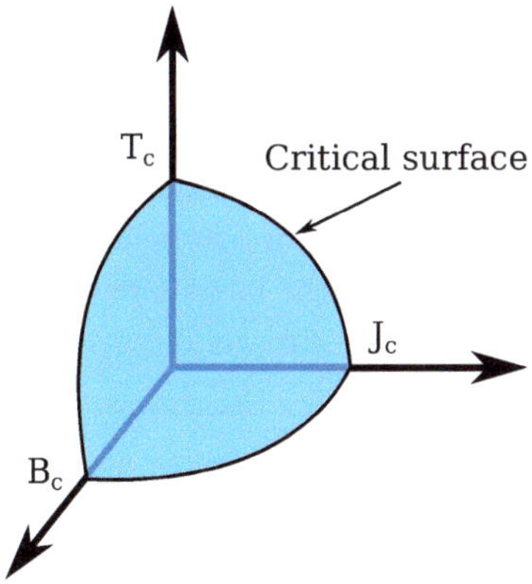

Figure 2.2: A superconductor will remain superconducting as long as it stays below its critical temperature, current density and magnetic flux density. The superconducting region is shown by the coloured area in the figure.

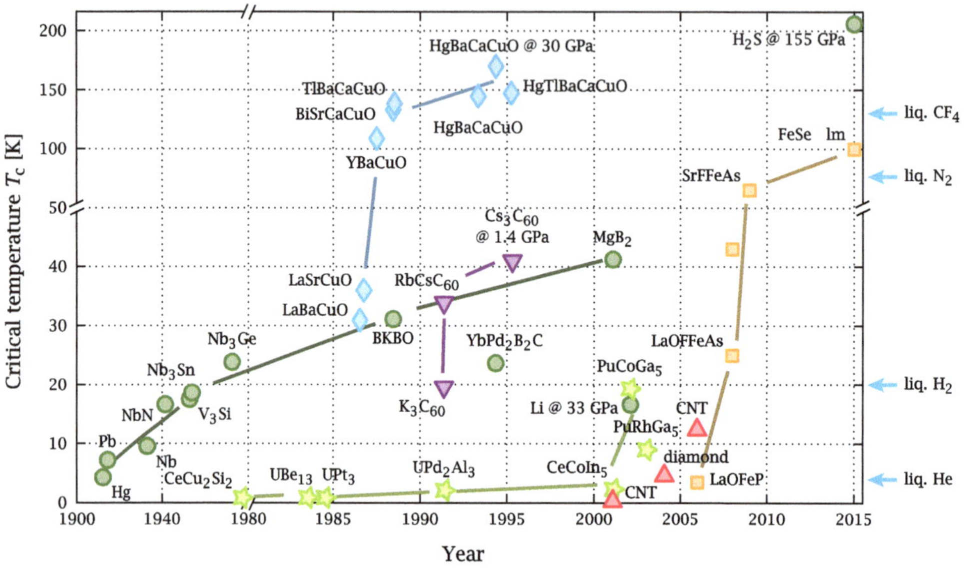

Figure 2.3: Discovery of various superconductors from 1911 to 2015 including their critical temperatures and the boiling temperature of common cryogenic liquids at atmospheric pressure [24].

Type I superconductors expel magnetic fields from inside the superconductor and the superconducting state is destroyed as soon as $B > B_c$. In type II superconductors, however, two distinct critical magnetic fields exist, B_{c1} and B_{c2}, where $B_{c1} \ll B_{c2}$. While $B < B_{c1}$, the magnetic field is expelled from the superconductor's interior, similar to that of type I superconductors. Whereas when $B_{c1} < B < B_{c2}$ the magnetic field penetrates the material in the form of flux vortices. This does not destroy the superconducting state immediately, the conductor remains superconducting until the B_{c2} magnetic field limit is reached.

Within HTS a further separation can also be made between first and second generation conductors, commonly referred to as 1G and 2G HTS, respectively. The former, 1G HTS were bismuth-based superconductors embedded in a silver matrix. The focus point of this research work, however, is the latter subgroup, hence in the following only second generation, high-temperature superconductors (2G HTS) of type II will be discussed.

2.2 HTS coated conductor tapes

This work focuses on the measurements, characterization and stability of HTS coated conductors and devices, more precisely $RE\mathrm{Ba_2Cu_3O_{7\text{-}x}}$ coated conductors (also simply called REBCO). Here *RE* stands for rare-earth and is commonly either yttrium or gadolinium. The respective superconductors are then referred to as $YBa_2Cu_3O_{7-x}$ or $GdBa_2Cu_3O_{7-x}$, or more commonly YBCO and GdBCO. These types of 2G HTS are also called "coated conductors" due to their structure and related manufacturing process.

2.2.1 Structure of the tapes

HTS coated conductors are manufactured as layers of materials on top of a substrate layer, such as illustrated in Figure 2.4. The manufacturing steps and technology varies between companies, however generally it begins with a metallic substrate, which is commonly Hastelloy® C276 with a thickness ranging in between 50 µm and 100 µm [27–30] and a width of 4 mm and 12 mm. On top of the substrate, buffer layers are deposited with a thickness of only a few hundred nanometres. These layers facilitate the growth and proper bonding of the superconducting YBCO or GDBCO layer, which is commonly 1 µm to 3 µm thick. Some manufacturers are, however, already providing superconducting layers of up to ~5 µm, which increases the I_c of their tapes substantially [29]. The next layer is 1 µm to 3 µm thick silver for protecting the superconducting layer and for finishing – if required – 10 µm to 100 µm thick metallic shunt layer, commonly copper. This custom metallic shunt finish is not strictly required and tapes that omit this shunt layer are often called "Ag-cap" or "stabilizer-free". HTS tapes with some form of metallic stabilizer on the other hand are referred to as "stabilized" or tapes with a "metallic shunt". This terminology will also be used throughout this work.

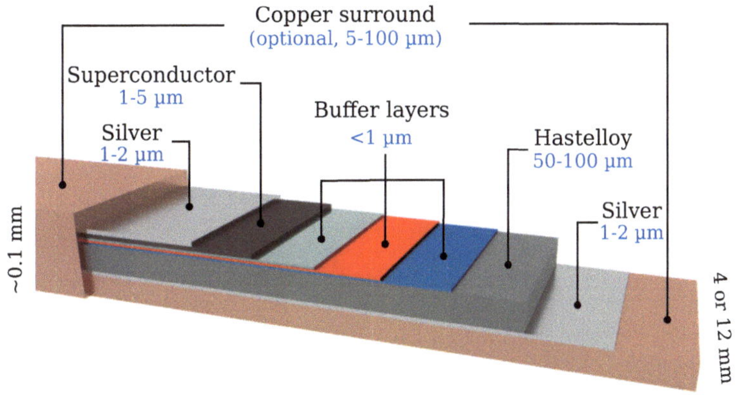

Figure 2.4: Typical architecture of a commercial, HTS coated conductor using Hastelloy® C276 as a substrate with surround copper stabilization. Note that the layers are not to scale.

It is important to mention that the layer thicknesses, the composition and number of buffer layers as well as the choice of the superconducting layer depend on the manufacturing process of the supplier. More information on this can be found at the respective manufacturer's website and product data sheets, for example [28–30].

2.2.2 Performance and applications

The critical current for commonly available 12 mm-wide HTS ranges between 300 A to 500 A at 77 K, in self-field (sf) [28, 30, 31]. This corresponds to engineering current densities, considering the whole thickness of the tape, of up to $4.2 \times 10^8\,\mathrm{A\,m^{-2}}$ [31].

While LTS can also provide current densities required for high field magnets, due to the $J_c(B)$ relationship a maximum field of about 15 T is commonly achievable [32] with the limits around 21 T for solenoids and 18 T for dipoles [33]. The additional benefit of HTS – apart from the higher allowable operating temperature – is the improved performance in magnetic fields. This combination of high current density in high magnetic fields make them an attractive candidate for a range of electrical applications, such as motors, generators, cables and high field magnets [3–8, 34]. As such, HTS can provide magnetic fields beyond 15 T by using an LTS magnet "outsert" generating a background field of 15 T and an HTS magnet "insert", such as in the 32 T all-superconducting magnet at the national high field magnet laboratory in Tallahassee, USA [32, 35–37] as well as several other projects worldwide [4]. It is important to note that albeit the current price of HTS is often still too high for commercial use, it is an enabling technology for magnetic fields above 45 T [38].

2.2.3 Anisotropy and magnetic fields

The thin yet relatively wide tape-like architecture gives rise to anisotropic material properties in 2G HTS tapes. One of the prominent effects of this architecture is the behaviour in magnetic fields. Generally speaking, magnetic fields perpendicular to the superconductor layer cause a larger reduction in critical current density than fields parallel to the superconducting layer. As such, $J_c(B, \Theta, T)$ can be defined for such coated conductors, where Θ is the angle between the actual field orientation and perpendicular. Here $0°$ and $90°$ mean a magnetic field perpendicular and parallel to the face of the tape, respectively. The angular field dependence of the critical current of a commercial, 12 mm-wide SuperOx tape is shown in Figure 2.5 at 77 K. This property of HTS tapes is especially important in HTS coils, where different parts of the coil experience fields of varying magnitude and orientation.

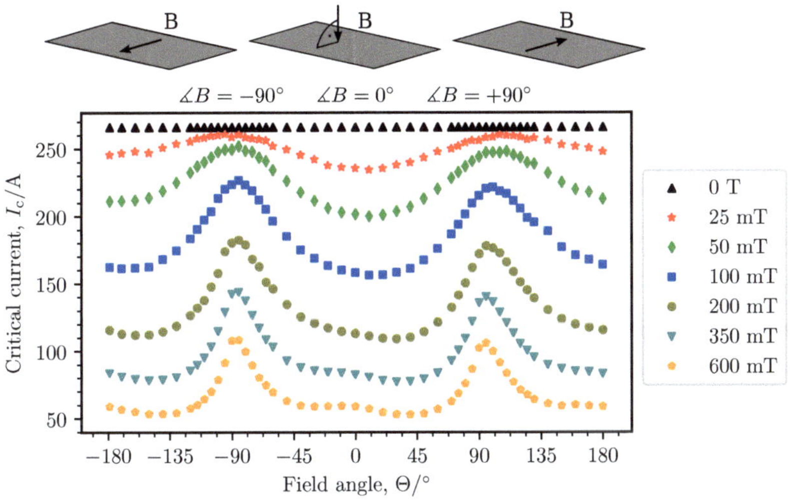

Figure 2.5: Dependence of the critical current on the orientation and amplitude of the applied magnetic field in a 12 mm-wide SuperOx HTS tape at 77 K. An angle of 0° represents a magnetic field perpendicular to the tape's face, whereas 90° corresponds to a field parallel to the plane of the tape.

2.2.4 Current-voltage characteristics

Since the amount of current a conductor can carry is limited by the critical current density, in practical application the need arises to experimentally determine a given conductor's critical current. In a superconductor, measuring a voltage drop V_{sc} between two points results in zero as long as the superconducting state exists. To describe the current versus voltage characteristics, the "power law" is commonly used for determining an electric field E_{sc} developed in a superconductor as

$$E_{sc} = E_c \left(\frac{|J|}{J_c} \right)^n , \tag{2.3}$$

where n describes the sharpness of the transition and is in the order of 30 to 80 for LTS and 10 to 40 for HTS and E_c is commonly $1\,\mu V\,cm^{-1}$ ($1 \times 10^{-4}\,V\,m^{-1}$) [22]. Consequently, the critical current of a given superconductor is determined at the point where $E_{sc} = E_c$. A typical E-J characteristic is illustrated in Figure 2.6, also showing the effect of the n-value.

The E_c criterion is merely a limit, which can be measured by accurate measurement devices. In the case of long length conductors, e.g. wound into the shape of a pancake or racetrack coil, an even smaller electric field criterion can be used, such as $0.1\,\mu V\,cm^{-1}$ ($1 \times 10^{-5}\,V\,m^{-1}$), provided that the electric noise in the measurement does not distort this reading. Yet again in other measurements with short, high amplitude

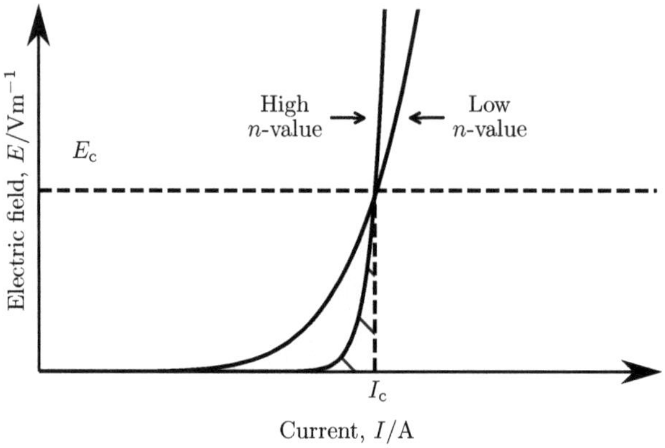

Figure 2.6: Definition of the critical current in superconductors using the E_c criterion and the importance of the n-value. While the conductors have an identical critical current, the one with a lower n-value will develop an electrical field sooner and hence incur I^2R losses as indicated by the shaded areas.

current pulses, a higher electric field criterion is used for claiming the change between superconducting and resistive state. This is necessary in fast measurements since the relatively slow nanovoltmeters are not suitable, and instead analogue to digital converters (A/D converters) are implemented. These allow a higher data acquisition speed at the cost of reduced sensitivity. In such experiments an electric field criterion of $1\,\mathrm{V\,m^{-1}}$ to $10\,\mathrm{V\,m^{-1}}$ is commonly used [18, 39, 40].

2.2.5 Losses in HTS tapes

While in direct current (dc) operation superconductors are truly dissipation-free, losses do arise when transient magnetic fields are present. The in-depth explanation of such effects is beyond the scope of this work, for detailed information on transient losses refer to [22, 41]. An approximation can be given as

$$P_{\text{loss}} = \begin{cases} 0, & \text{when } I < I_c \\ > 0, & \text{when } B = B(t) \\ > 0, & \text{when } i(t) = I\sin(\omega t). \end{cases} \tag{2.4}$$

Here P_{loss} is the power loss, I represents a dc current, $i(t)$ represents an alternating (ac) current and ω is the angular frequency.

Hysteresis losses

As mentioned in Section 2.1.3, in HTS tapes the magnetic field lines penetrate the conductor in so-called flux vortices. As long as the magnetic field does not change and the conductor is in superconducting state, these vortices are pinned to material defects and cause no dissipation. However, in an alternating magnetic field the vortices have to re-arrange continuously and do work, giving rise to hysteresis losses. Based on the origin of these losses one may talk about transport losses or magnetization losses. The former is caused by the transport current in the conductor and the related self-field, whereas the latter is induced by an external magnetic field.

Resistive losses

Resistive losses arise in superconductors when the normal transition begins. As visible in Figure 2.6, the conductor with a lower n-value would incur $E \cdot J$ losses even at current levels below the critical current. Above the critical current the losses are determined by the compound electrical resistance of the conductor, including the electrical resistance of the normal metal matrix.

Eddy current losses

Eddy current losses occur in the normal metal parts of 2G HTS tapes, such as the copper stabilization, in the case of changing magnetic fields as $P_{\text{eddy}} \propto f, B^2$. Due to low magnetic fields discussed in this work, eddy current losses can be neglected.

2.3 Stability of HTS tapes

Superconductors are generally bad electrical conductors at room temperature, hence it is highly desired to keep them superconducting at all times when in operation. Due to some disturbance, however, it can happen that a superconductor experiences a local reduction in J_c, such as illustrated in Figure 2.7. This may be caused by a locally increased magnetic field, a material defect or an additional local heat load. Such a local reduction in critical current density can move the local operating point of the superconductor closer and even above the critical surface (Figure 2.2) and heating occurs in the form of Joule losses over a small volume. If such a hot spot is undetected, the continued heat input will cause further heating while at the same time the heat propagates, pushing the operating point of an increasing volume potentially above the critical surface. Provided that the rate of heating due to the Joule loss equals the rate of cooling (e.g. in a cryogenic bath), the superconductor may still be in a cryostable state. And while the situation is no longer considered truly lossless, the superconductor will not be damaged and remains operational. Based on this definition a minimum propagating zone (MPZ) can be calculated, defined as the minimum length of a hot zone that will cause it to expand and lead to a thermal runaway. Any length smaller than this will not produce enough heating to cause a thermal runaway and will disappear due to the cooling. Any larger zone, however, will cause a thermal runaway, also referred to as a quench, which is a rapid and abrupt transfer of the superconductor from superconducting to normal conducting state. The energy required to cause an MPZ is referred to as the minimum quench energy (MQE) and is an often-used characteristic describing the amount of energy input a superconductor can absorb before quenching. To calculate the MPZ length l_{mpz} in a flat HTS tape submerged in a cryogenic bath a simplified equation can be given by equating the heat generation with the heat transfer away from the zone in both directions as well as convective cooling as

$$\underbrace{J^2 \rho_{\mathrm{cd}} l_{\mathrm{mpz}} A_{\mathrm{t}}}_{\text{Heating}} = \underbrace{h(T - T_{\mathrm{op}}) S_{\mathrm{t}} l_{\mathrm{mpz}}}_{\text{Convection}} + \underbrace{2 \frac{\kappa_{\mathrm{cd}} A_{\mathrm{t}} (T - T_{\mathrm{op}})}{l_{\mathrm{mpz}}}}_{\text{Conduction}}. \tag{2.5}$$

Here ρ_{cd} is the compound electrical resistivity, A_{t} the cross-sectional area of the HTS tape, h the heat transfer coefficient to the ambient medium, T is the temperature of the normal zone, T_{op} is the ambient (operating) temperature, S_{t} is the perimeter of the

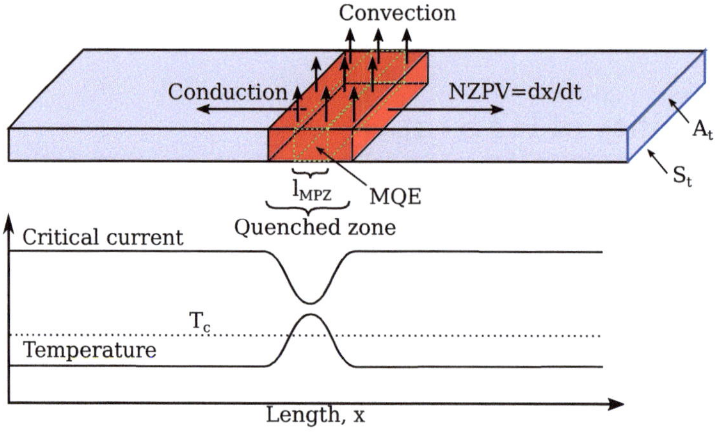

Figure 2.7: Concepts of quench, normal zone propagation, minimum quench energy and minimum propagation length in a HTS coated conductor cooled in a cryogenic bath. The arrows represent heat transfer by different mechanisms, whereas the bottom plot illustrates the critical current density and temperature profile of the tape.

conductor and κ_{cd} is the compound heat conductivity of the conductor. The equation also assumes that the conductor is at the intial temperature of T_{op}. Then expressing the MPZ gives [42]

$$l_{\mathrm{mpz}} = \sqrt{\frac{2\kappa_{\mathrm{cd}} A_{\mathrm{t}} (T - T_{\mathrm{op}})}{J^2 \rho_{\mathrm{cd}} - \frac{h S_{\mathrm{t}}}{A_{\mathrm{t}}} (T - T_{\mathrm{op}})}}. \tag{2.6}$$

The speed at which this thermal wave front propagates is called the normal zone propagation velocity (NZPV). As indicated by eq. 2.5 this speed will depend on a range of factors, including the operating current, temperature margin (difference between operating and the critical temperature), as well as the heat conductivity and capacity.

LTS applications are often operated at 4.2 K, where the heat capacity of metals is significantly lower than at room temperature and the temperature margin of LTS is low. In HTS, on the other hand, when operated at 77 K in liquid nitrogen the heat capacity of common metals is two orders of magnitude higher. It is also not uncommon to operate an HTS application at lower temperatures to yield a higher operating current density [37, 43]. This further increases the temperature margin and consequently the amount of energy that can be absorbed. Assuming adiabatic conditions during a fast quench and neglecting heat transfer to a liquid nitrogen or helium bath as well as thermal conduction, the minimum quench energy is described by [22]

$$Q_{\mathrm{MQE}} = \int_{T_{\mathrm{op}}}^{T_{\mathrm{cs}}(I_{\mathrm{op}})} C_{\mathrm{cd}}(T) dT. \tag{2.7}$$

Here C_{cd} is the compound heat capacity of a unit volume conductor and T_{cs} represents the current sharing temperature, defined as the point where current starts flowing in the normal metal matrix. This is commonly defined as [22]

$$T_{\mathrm{cs}} = T_{\mathrm{op}} + (T_{\mathrm{c}} - T_{\mathrm{op}})(1 - \frac{I}{I_{\mathrm{c,\,op}}}). \tag{2.8}$$

Here $I_{\mathrm{c,\,op}}$ is the critical current of the conductor at the operating temperature. A comparison of MQE for LTS and HTS composite conductors is given in Table 2.1 [22]. It is clearly visible that HTS have a 2 to 3 order of magnitude larger MQE than LTS.

Table 2.1: Common values of T_{op}, ΔT_{op} and Q_{MQE} for LTS and HTS [22].

LTS			HTS		
T_{op} (K)	$\Delta T_{op}(I_{op})$ (K)	Q_{MQE} ($J\,cm^{-3}$)	T_{op} (K)	$\Delta T_{op}(I_{op})$ (K)	Q_{MQE} ($J\,cm^{-3}$)
2.5	0.3	1.2×10^{-4}	4.2	25	1.6
4.2	0.5	0.6×10^{-3}	10	20	1.8
4.2	2	4.3×10^{-3}	30	10	3.7
10	1	9×10^{-3}	70	5	8.1

2.3.1 Normal zone propagation

Reduced heat capacity, temperature margin and an increased thermal conductivity of LTS operated at low temperatures causes local hot spots to spread rapidly in case of a quench, generally in the order of magnitude of several $m\,s^{-1}$ [14, 44] with sources reporting speeds of up to $km\,s^{-1}$ [45, 46]. Hence if an LTS cable or device quenches, the heating power is quickly distributed over a longer length (and larger mass) of material and the current is limited by the high resistance. HTS applications, however, are inherently more stable and resilient towards disturbances due to the increased heat capacities as well as temperature margins. On the downside, in the case of a local hot spot, they propagate significantly slower than in the case of LTS devices. Therefore, the characteristic NZPVs in HTS are in the range of $cm\,s^{-1}$ [12, 13, 15, 16] with ongoing work to speed up the propagation into the $m\,s^{-1}$ range [17, 18, 47].

The primary benefit of a metallic stabilization (Section 2.2.1) in HTS is to provide an alternative path for the current in the case of a quench and to prevent damage to the superconducting as well as the thin silver layer. At the same time, however, it increases the thermal mass and thereby the stability as well (see eq. 2.7), and reduces the NZPV [48] in the case of a quench. And while an increased stability is definitely beneficial for most applications, in case of a quench a faster NZPV is considered as a protection mechanism for the conductors and it is mostly well-seen, even sought after [15]. For these reasons the quench and normal zone propagation are an actively researched area of superconductivity [49].

2.3.2 Minimum quench energy measurement methods

For measuring the MQE, commonly the HTS tapes are cooled down and a steady dc current is applied such that $I_{\mathrm{op}} < I_{\mathrm{c}}$. Then, a small heater mounted on the surface of the tape is pulsed with a short current pulse. The generated heat reduces locally the I_{c} of the conductor and (provided that enough energy was supplied) a quench is caused. The pulse energy level where the quench first appears is usually called the MQE. In the same measurement, often an NZPV can also be calculated for sub-critical current levels. Nevertheless, there is some degree of uncertainty in measurements with heater pulses due to both heat transfer rates into the superconductor as well as the heat capacity of the heaters themselves [15]. Furthermore, the size of the heater directly determines the area absorbing the supplied energy, where e.g. Pelegrin et al. [15] used a $4\,\mathrm{mm} \times 4\,\mathrm{mm} \times 0.2\,\mathrm{mm}$ graphite-based heater.

Referring to Table 2.1, as a summary it can be stated that while LTS have an MQE in the order of $\mu\mathrm{J\,cm^{-3}}$ [44] resulting in a vastly faster NZPV, HTS have an inherently higher heat capacity and temperature margin, resulting in a larger MQE, in the order of $\mathrm{mJ\,cm^{-3}}$ to $\mathrm{J\,cm^{-3}}$ [15, 16, 50], but a significantly reduced NZPV.

2.3.3 NZPV measurement methods

The state-of-the-art method for calculating the NZPV in HTS conductors is to time the appearance of an electrical signal over short distances of a conductor [15–18, 51–53]. An HTS probe is placed in a sample holder with several voltage taps aligned over the surface at predefined distances, commonly ~5 mm and similar to that illustrated in Figure 2.8. The tape's critical current density is reduced over a single point by the use of either a heater, magnetic or a mechanical defect. The sample is then cooled down in a liquid nitrogen bath and current is applied in form of a short pulse. A quench is nucleated at the position of reduced J_{c} and by recording the developing electric field between the voltage taps, the "flight" of a normal zone, NZPV, can be calculated.

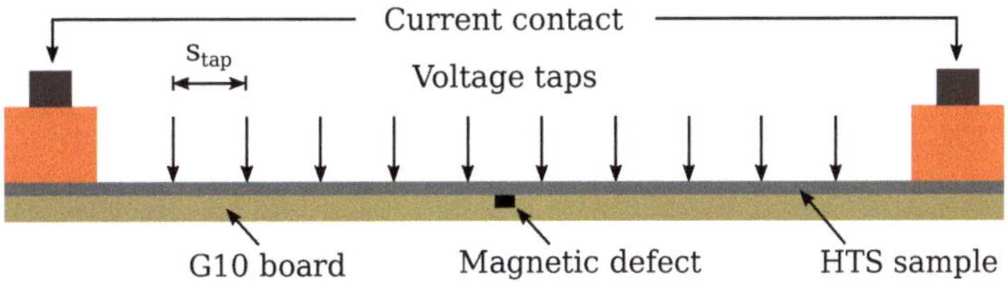

Figure 2.8: Cross-section drawing of a common experimental setup used to measure the NZPV, showing several voltage tapes positioned in precise distances.

2.3.4 Quench detection and protection

Apart from measuring voltage signals, other non-intrusive methods have also been researched for quench detection and localization, some of which are also suitable for determining the NZPV. Such are, for example, the use of optical fibres or acoustic signals in a conductor.

The use of commercial optical fibres (embedded or co-wound in the conductor) for quench detection and localization has been already investigated and validated for the use in HTS applications [54–56]. The method may be implemented using different principles, but most of the recent applications focus on Rayleigh scattering. The slight changes in temperature or strain in the optical fibre, caused e.g. by a hot spot, change the scattering of light in the fibre, which can then be detected and compared with reference signals.

Acoustic methods are also already available for detecting and locating quench in superconductors. Passive acoustic emission can be used to forecast a quench occurrence in LTS during training [57, 58]. An active technique for detecting and localizing quench has also been validated by both simulations and experiments [59, 60]. In this method acoustic pulses are sent across conductors and the change caused by hot spots in the measured signal is used for localizing the warmer region with a detection accuracy of 1 K.

Another electrical method proposed the use of magnetic field sensors to measure changes in the magnetic field caused by current redistribution between strands of an HTS tape [61]. Although the method requires the pre-processing of HTS to have (at least) two separate strands, the technique can in principle forecast a quench before any voltage signal could be measured.

2.4 HTS coils

As discussed in Section 2.3, due to the orders-of-magnitude larger heat capacity of HTS, their stability is vastly superior to that of LTS counterparts. This applies to magnets and coils alike and as such, premature quench due to small, local disturbances – like in the case of LTS – is no longer a concern in HTS [22]. Nevertheless, similar to single conductors, coils wound from HTS may experience larger disturbances, due to issues with cooling, overcurrent, local material defects or ac losses. If undetected, these can still lead to a local temperature rise that can permanently reduce the I_c of the coil. Therefore it is of high importance to detect potential local hot spots and thermal runaways, preferably even before they would occur. This means that a system for detecting a quench is required together with a method to mitigate and protect the coil from permanent damage.

2.4.1 Quench detection and protection

Several active and passive methods exist for the detection and protection of coils in the case of quench. Quench detection is often done using the bridge-balance method [62], where a voltage signal is monitored between a potentiometer and a coil centre tap. However, other methods for detecting a quench are also possible, such as the described optical and acoustic methods in Section 2.3.3 or co-winding the HTS coil with an insulated, metallic tape. This latter approach couples the two systems magnetically and effectively cancels out the inductive coil voltage when measured between the two windings, leading to a higher detection sensitivity [62].

In all cases, a signal is measured (voltage, acoustic or optical) and compared to a threshold value determined for the specific application. When this threshold is exceeded and a quench is detected by one of the methods, a quench protection system is engaged. Common, active quench protection systems include using a dump resistor to discharge the coil energy [63–66], resistive heaters to create a larger normal zone [32, 35, 49, 66], coupling-loss-induced bulk heating (CLIQ) [67, 68] or ac loss-induced heating. However, apart from active quench protection methods, passive approaches exist as well for HTS coils, such as the no-insulation winding technique as introduced by Hahn et al. [69].

2.4.2 Non-insulated coils

Figure 2.9 illustrates the difference between an insulated and a non-insulated (NI) coil wound from 2G HTS tapes. In an insulated coil (Figure 2.9a), the spiral path of the conductor is the only path where current may flow and the coil can be described by an equivalent circuit as in Figure 2.10a. The main difference to an ordinary coil is that during the HTS tapes' superconducting phase in direct current operation there is no power dissipation. Nevertheless, a local disturbance or defect in the material may still cause local dissipation, which can lead to local heating and a thermal runaway inside the coil.

Contrary to this, in an NI coil (Figure 2.9b), the insulation between the turns is omitted. This creates an alternative electrical path, where current may also flow in the transverse direction as illustrated in the equivalent circuit in Figure 2.10b. The benefit of this is an increased stability in case of a local disturbance, in which the coil current may (partially) bypass the section with a locally reduced J_c. While the path across the turn-to-turn resistances is resistive, and hence incurs Joule heating, the superconductor is protected from overheating and potentially a local quench. The self-protection mechanism has already been reported and investigated in several works both experimentally and in simulations [70–74].

Charging time constant

The universal time constant in an LR circuit is defined as

$$\tau = \frac{L}{R}, \tag{2.9}$$

where L is the inductance and R the resistance of the coil. This is the time required for the voltage or current to reach

$$1 - \frac{1}{\mathrm{e}} \cong 0.63 \tag{2.10}$$

of the final value after a switch has been closed in a voltage-driven circuit and current started flowing. In superconductor coils, however, the voltage (or current) is not applied in a step, but as a slow ramp. In this case the rise of current is controlled and the measured voltage depends on the inductance and the ramping rate,

$$V_{\mathrm{coil}} = L\frac{\mathrm{d}I_{\mathrm{s}}}{\mathrm{d}t} + I_{\mathrm{s}}R_{\mathrm{s}} = I_{\mathrm{r}}R_{\mathrm{r}}. \tag{2.11}$$

Here R_{r} is the coil resistance across the transverse path (Figure 2.9b), R_{s} is the resistance along the spiral path (resistance of the superconductor), I_{r} is the current flowing in the radial path and I_{s} is the current in the spiral path. Furthermore, in an NI superconducting coil the meaning of τ is different as currents can flow across the turn-to-turn contact resistances as well, effectively bypassing the coil's inductance, depending on the ramping rate. As a consequence, τ depends on the current ramping rate. Therefore, it is also different in case of a sudden discharge, which can be interpreted as a step (infinitely fast) change in the coil current. Hence one can differentiate between charging time constant τ_{up} and a discharge time constant τ_{down}. Kim et al. defined the charging time constant of a coil, as the time it takes for the magnetic field to reach 63 % of the difference between the maximum field and the field where the ramping has finished, measured from the time when the current reached the target value [75]. This is illustrated in Figure 2.11a by the dashed area, where the magnetic field changes from B_0 to B_1 and

$$\tau_{\mathrm{up}} = t_1|_{B_1} - t_0|_{B_0}. \tag{2.12}$$

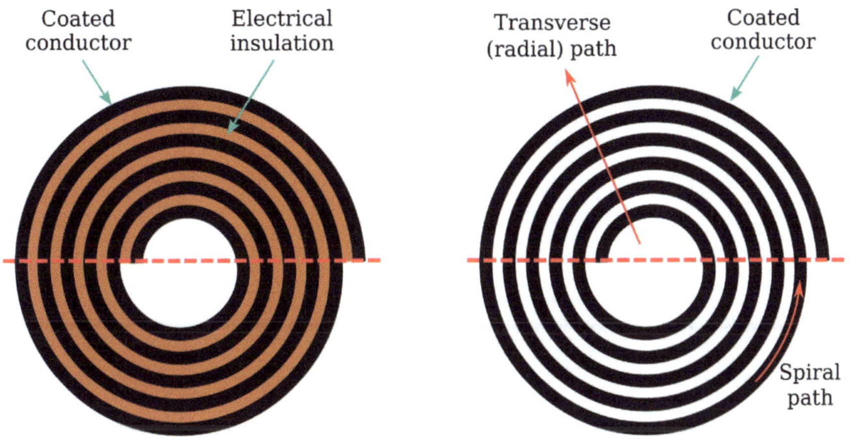

(a) Insulated coil, top view.

(b) Non-insulated coil, top view.

(c) Insulated coil, cross-section.

(d) Non-insulated coil, cross-section.

Figure 2.9: Concept drawing of an insulated coil (a) and a non-insulated coil (b), with corresponding cross-sections shown across the red dashed line in (c) and (d), respectively. The red arrows indicate possible current paths.

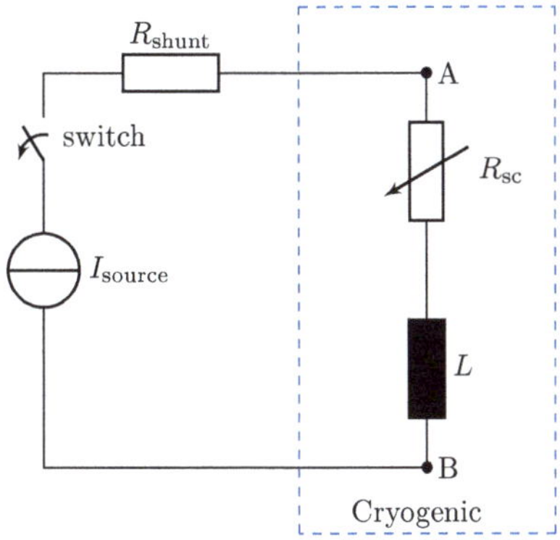

(a)

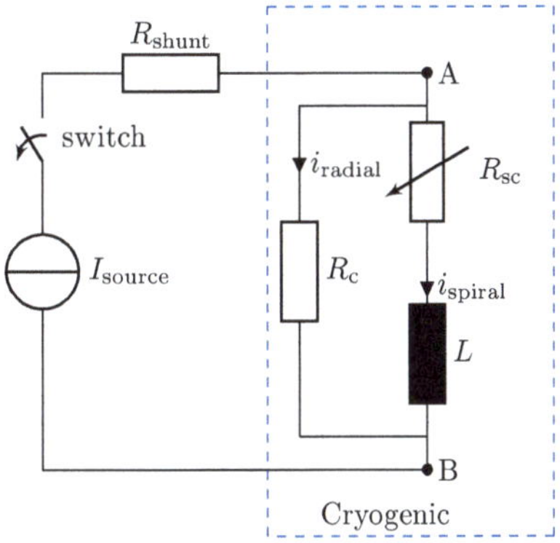

(b)

Figure 2.10: Simplified circuit diagrams for (a) an insulated HTS coil and (b) a non-insulated HTS coil also showing the cryogenic environment. The coil voltage is measured between points A and B.

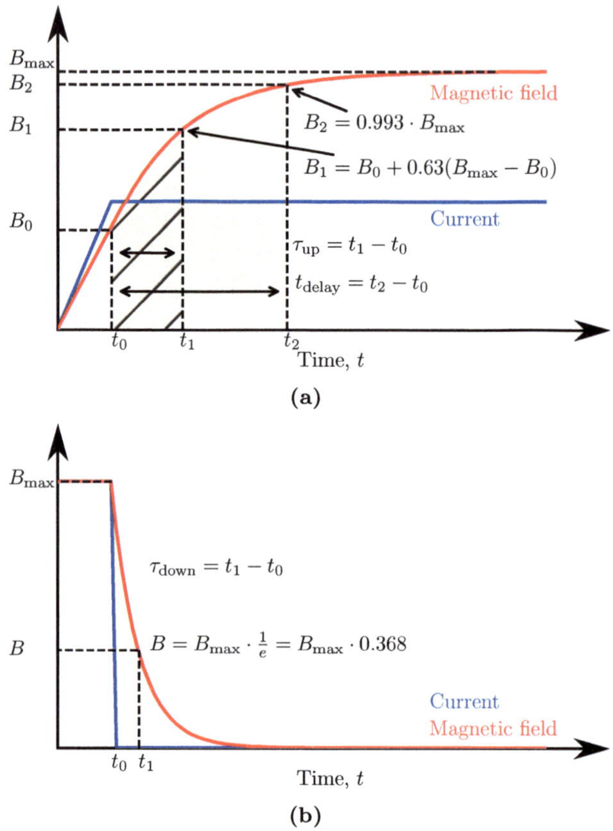

Figure 2.11: Concepts of various time constants. (a) shows the charging time constant τ_{up} in the hatched area and the charging delay t_{delay} in the coloured area. (b) shows the discharge time constants τ_{down} in the coloured area. The features of the plots are exaggerated for visualization.

Discharge time constant

An NI coil discharge time constant τ_{down} as shown in Figure 2.11b can be obtained by ramping the coil to a specific current level and opening a switch in the circuit (refer to Figure 2.10). In this case, without using any dump resistor, the coil's energy is discharged across the turn-to-turn resistances. From the moment the switch was opened, the decay of the magnetic field is measured until it reaches

$$\frac{1}{e} \cong 0.368 \tag{2.13}$$

of its initial value. The discharge time constant is then

$$\tau_{down} = t_1|_{B_{max}} - t_0|_{B_{max} \cdot 0.368}, \tag{2.14}$$

according to Figure 2.11b. Kim et al. showed that the discharge time constant does not depend on the current level from where the coil is discharged in experiments including 20 A, 40 A, 60 A and 80 A current plateaus [75]. They also concluded that the characteristic coil resistance R_c changes when calculated using τ_{up} in eq. 2.9, since the final current amplitude changes the τ_{up} itself when the ramping rate was held constant at $2\,\mathrm{A\,s^{-1}}$.

Charging delay

In an NI coil the current may not fully follow the spiral path due to the lack of insulation between the turns. As such the total current in the coil is

$$I_{coil} = I_s + I_r, \tag{2.15}$$

where $I_{coil} = I_{source}$ is the overall driving current of the circuit. Since the magnetic field is predominantly generated by the current in the spiral path (I_s), from e.q. 2.15 and e.q. 2.11 follow that during the transient ramping phase the magnetic field will "lag" the net current supplied to the coil. And while the magnetic field does stabilize eventually, it results in a certain charging delay. This is defined as the time it takes for the coil to reach 99.3 % of its final magnetic field at the centre of the coil, with respect to the time when the ramping has finished [76, 77]. This is illustrated in Figure 2.11a, where the 99.3 % level originates from the 5τ (five time constants) and the charging delay is defined as

$$t_{delay} = t|_{B_{max}(1-\frac{1}{e^5})} - t|_{\frac{dI}{dt}=0}. \tag{2.16}$$

The charging time delay has a practical importance, especially in larger magnets with large inductances, as it adds a considerable time to the total ramping duration [77].

Turn-to-turn resistance

Another important aspect of NI coils is the resistance between the turns of the coil often given as the turn-to-turn contact resistivity ρ_{ct}, defined as the contact resistance multiplied by the contact area and having units of $\mu\Omega\,\mathrm{cm}^2$. This is commonly calculated from the characteristic coil resistance R_c, which is the sum of all turn-to-turn resistances [76, 78, 79],

$$R_c = \sum_{i=1}^{N_t-1} R_i = \sum_{i=1}^{N_t-1} \frac{\rho_{ct}}{2\pi r_i w_d}. \tag{2.17}$$

Here N_t is the total number of turns, r_i and R_i are the radius and the resistance across the ith turn and w_d is the conductor width. In practice, R_c is calculated from the coil inductance and the time constant, τ, according to eq. 2.9. However, as it will become apparent in Chapter 5, the time constant not only changes during charging and discharging, but it also varies depending on the ramping rate. It follows then that the characteristic coil resistance must also change as a function of the time constant and consequently the ramping rate. For this reason, commonly the discharge time constant is used in calculating R_c, since this does not depend on the applied current.

According to eq 2.17 the characteristic coil resistance is directly related to the turn-to-turn contact resistivity. Small resistivity values lead to a small coil resistance and hence according to eq 2.9 a large τ, causing long charging delays. An approach to mitigate this issue is the artificial increase of contact resistivity by means of metal co-winding or various surface treatment techniques.

It is known that the contact resistivity strongly depends on the winding tension, contact surface quality (roughness, oxidation), tape width and copper stabilization thickness as well as the operating temperature [80–82]. Therefore it is difficult to state common values, however 70 µΩ cm² [76] is commonly referenced in literature and used in numerical simulations [73, 78, 83–85]. A broader range of resistivity values can be given from 5 µΩ cm² until 9.8 mΩ cm² [74, 76, 78, 80–82, 86, 87] depending on the previously mentioned features. These values are commonly calculated indirectly by substituting eq. 2.9 into eq. 2.17 and solving for ρ_{ct},

$$\rho_{ct} = \frac{\frac{L}{\tau_{down}}}{\sum_{i=1}^{N_t-1} \frac{1}{2\pi r_i w_d}}. \tag{2.18}$$

Lu et al. [81, 82] carried out extensive measurements into surface contact resistivity between REBCO tapes using different surface cleaning methods as well as temperature and load cycling. They showed that the surface contact resistivity decreases with increasing load, but initially increases with an increasing number of load cycles. The authors attributed this initial increase in surface resistivity to cryogenic work hardening of copper at 77 K and measured resistivity values between 26 µΩ cm² to 100 µΩ cm². In a later work they showed, however, that after a peak resistivity, additional load cycles reduced ρ_{ct} [82]. While this effect is less pronounced at 77 K (up to a factor of ~6), at 4.2 K a large number of load or thermal cycles may cause a surface resistivity several orders of magnitude lower than initially.

3 A high-speed thermal imaging method for cryogenic temperatures

The first part of this chapter is dedicated to briefly discussing for comparison purposes two state-of-the-art thermal imaging methods: infrared thermal imaging and fluorescent micro-thermographic imaging. The second part of the chapter discusses the newly developed high-speed fluorescent thermal imaging method for quench investigation in superconductors. The principles of fluorescent thermal imaging are explained in sufficient detail to support the understanding of measurement results presented in later chapters.

3.1 State-of-the-art thermal imaging methods

3.1.1 Infrared thermal imaging

A well-known thermal imaging method is infrared thermography (IRT), which uses the black-body radiation of objects to determine their temperature. The term "black body" refers to an ideal body in physics, which absorbs all incident energy. With α as the absorption, τ the transmissivity and r the reflectivity the total radiation at a surface is $1 = \tau + \alpha + r$, which in the case of a black body becomes $\alpha = 1$. Furthermore, an ideal black body is also considered as a perfect emitter and has an emissivity of $\epsilon = 1$. The emissivity of real objects is below this theoretical maximum and it is hence used as a percentage of a black body's. Black-body radiation is an electromagnetic radiation emitted by all objects above absolute zero temperature and it is proportional to the fourth power of the temperature

$$P = A \cdot \sigma \cdot \epsilon \cdot T^4. \tag{3.1}$$

Here P is the radiated power, A is the surface area, σ is the Stefan-Boltzmann constant, and T is the temperature in kelvin. This implies that while "hot" bodies radiate a lot of energy, "cold" objects radiate significantly less. This constitutes one of the significant limitations of the IRT method. As the temperature of the object to be measured reduces, the detectable energy emission decreases with the fourth power. Therefore, when measuring objects at low temperatures, where the environment temperature is significantly higher, can result in a considerably low signal. For this reason most IRT cameras have an approximate minimum temperature limit of −20 °C to −50 °C.

The wavelength of the black body radiation also depends on the temperature as defined by Wien's displacement law

$$\lambda_{\text{peak}} = \frac{b}{T} \tag{3.2}$$

where λ_{peak} is the peak emission wavelength in metres and b is Wien's displacement constant. While extremely hot objects can emit light in the visible spectrum, mostly when measuring non-astronomical objects the emitted radiation falls in the infrared (IR) spectrum of light, hence the term "infrared thermal imaging". A plot of the electromagnetic spectrum between 250 nm and 1 µm is shown in Figure 3.1 for reference.

By substituting values into equation 3.2 it can be calculated that for example the human body at a temperature of 37 °C would radiate energy at a peak wavelength of 9.34 µm.

From eq. 3.1–3.2 it can be seen that for relatively low temperatures the wavelength of the black body radiation falls always in the IR spectrum. While this radiation is invisible for the naked eye (Figure 3.1), cameras can be designed to be sensitive to this particular spectrum of light. Such infrared (or thermographic) cameras can

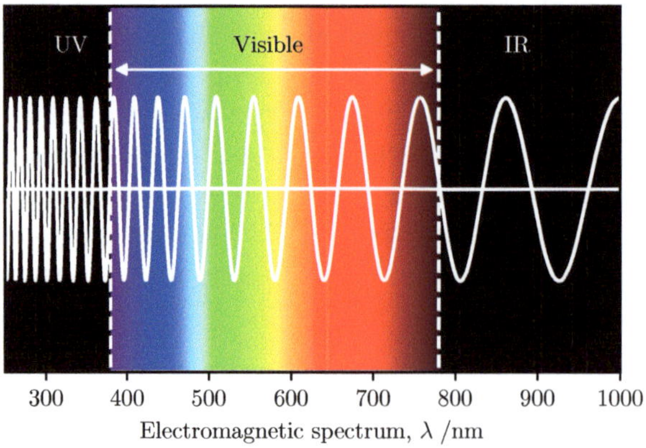

Figure 3.1: Representation of the electromagnetic spectrum between wavelengths of 250 nm (near ultraviolet, UV) and 1 µm (near-infrared, IR). The spectrum visible for the human eye between approximately 380 nm and 780 nm is highlighted and shown with the colours matching the wavelengths.

record the electromagnetic radiation of an object and use post-processing to find its temperature. However, since the recorded radiation is the function of the object's surface emissivity, standard values for different surfaces have to be empirically programmed into the post-processing algorithm. The method is also known to perform rather poorly when imaging surfaces with high reflectivity (such as metals) due to the excessive reflection [88]. When imaging such surfaces it may be necessary to artificially reduce the reflectivity and increase the emissivity by e.g. applying a black coating on the surface.

Apart from the useful temperature range of such an imaging technique another difficulty is related to the fundamentals of its operation, namely the wavelength of the recorded radiation. Since this falls in the infrared band of light, it limits the maximum spatial resolution (minimum resolvable distance between two points) obtainable to [89]

$$\text{Resolution} = \frac{0.61\lambda}{N.A.}, \tag{3.3}$$

where N.A. is the numerical aperture of the imaging device, e.g. microscope. Hence with an increasing wavelength the minimum resolution gets worse and as such the IRT method's resolution is limited by the wavelength of the black body radiation itself.

To summarize, IR thermal imaging works on the basic principles of black body radiation and hence does not require any coating or direct contact with the object to be measured, in most situations. Nevertheless, due to the long wavelength and low amount of energy radiation, IR camera sensors have to be manufactured with increasing efficiency to measure at low temperatures that quickly becomes challenging. The maximum spatial resolution is also limited by the long wavelength of the radiation, which can be limiting for microscopy applications.

3.1.2 Fluorescent micro-thermographic imaging

Fluorescent micro-thermographic imaging (FMI) is a method developed in 1982 by Kolodner et al. [90, 91] as a temperature measurement technique addressing the shortcomings of the IRT method. Its goal was to provide high spatial and thermal resolution, using the light intensity decay of fluorescence in rare-earth compounds with changing temperatures. Contrary to infrared thermal imaging, this method does not use the black body radiation emitted directly by an object, but relies on the photoluminescence of an applied coating. The basic principle is that if the fluorescence has a well-known temperature dependence, that is the light intensity varies predictably based on the temperature, then after a careful calibration measurement the temperature of an object can be calculated from the light intensity emanating from the fluorescent coating. It was demonstrated on MOSFETs in a packaged integrated circuit that such a method can have a temperature accuracy of 0.01 °C and a spatial resolution of 15 µm. A year later, a temperature accuracy of 0.08 °C and a spatial resolution of 0.7 µm were reported. Haugen et al. [92, 93] then used FMI for the first time to image microscopic superconducting bridges carrying small amounts of current and demonstrated that measurements are possible in superconductors down to 4.2 K in an experimental setup shown in Figure 3.2.

While the original concept uses solely the light intensity of a photoluminescent coating, Nara et al. [94] and Ishiyama et al. [95] used instead the colour change of a mix of different rare-earths instead of a single fluorophore. By the use of a mix of paints with different emission colours and temperature dependencies, the colour change with respect to temperature was calibrated and calculated. As the emission light intensity varies differently in the case of the two different paints, a colour shift can be observed and allowed the authors to reduce measurement error caused by the absorption of light along the path. Since in this method the actual colour change is important, it requires the use of either a spectrometer, resulting in only a point reading, or a colour camera. While this method provides quantitative temperature readings, in the current implementation it was only demonstrated for measuring thermal zones propagating over many seconds.

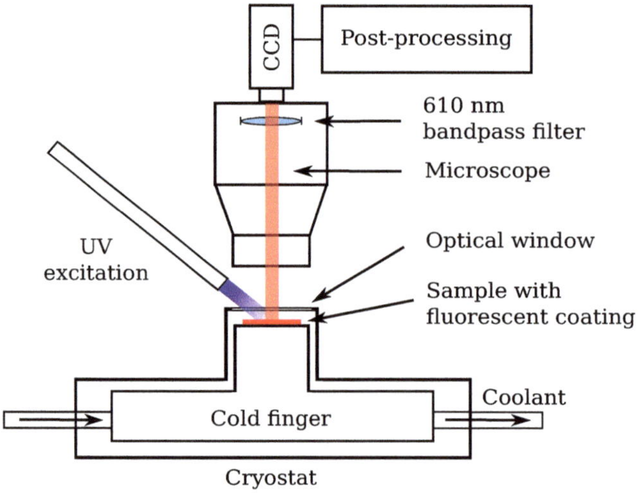

Figure 3.2: Experimental setup used by Haugen et al. [92, 93] for fluorescent micro-thermographic imaging until 4 K for imaging microscopic superconducting bridges.

3.2 High-speed fluorescent thermal imaging for quench analysis

This work aimed at the further development of the already existing fluorescent micro-thermographic imaging technique to make it usable at high speeds for analysing quench and measuring normal zone propagation in HTS. The capability of mapping the temperature distribution over a 2D surface at thousands of frames per second over a liquid nitrogen-immersed HTS tape, cable or coil constitutes the main novelty of the method compared to previous research. Although the method provides a fine temporal resolution in the microsecond range, it is not referred to as a "real-time" system, since the post-processing of the data is done separately. Hence it is rather aimed at the analysis and investigation of the quench behaviour itself and optionally also the recovery afterwards, in its currently introduced form.

The measurement principle described in Section 3.1.2 was adapted to measuring transient effects, the quench of a superconductor tape, at cryogenic temperatures. The finished measurement setup used for measuring HTS tapes is illustrated in Figure 3.3. A more detailed description of the assembly as well as the measurement procedure is given in Chapter 4.

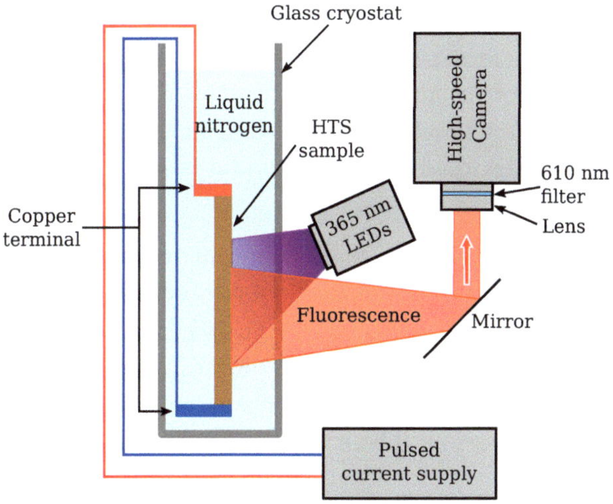

Figure 3.3: Cross-section schematic view of the experimental assembly. The sample is placed in a double-walled glass cryostat and submerged in liquid nitrogen. The UV LED array excites the sample at an oblique angle and the high-speed camera records the fluorescent light emission through a mirror due to the experimental arrangement.

3.2.1 The basic principles of photoluminescence

Discussing photoluminescence in greater detail is beyond the scope of this work and therefore only a brief introduction is given to aid the understanding of the high-speed fluorescent thermal imaging method. For a deeper understanding of these phenomena sources such as [96–98] should be consulted.

Fluorescence and phosphorescence are types of photoluminescence in which a material absorbs incoming electromagnetic radiation and re-emits it at another, usually higher, wavelength. In a simplified manner, the difference between fluorescence and phosphorescence is the duration of the delay between excitation and emission, however this can be led back to how the relaxation happens in different molecules. The process starts with the absorption of photons, whose energy is used to elevate the photoluminescent molecule from their ground state to some higher, excited state, as shown in Figure 3.4. At this excited state the molecule loses excess vibrational energy and returns to the lowest vibrational state of the excited state. Afterwards the fluorescent molecule can return to the the ground state by emitting energy in the form of photons, fluorescent light. The relaxation from the excited to the ground state is fast, in the order of 10^{-9} s to 10^{-7} s [96], meaning that if the excitation light source is terminated, the fluorescent light is practically immediately stopped as well. In phosphorescence, however, molecules are excited into a triplet state, as predicted by quantum theory. The relaxation from the triplet state back to the ground state is difficult and hence the process takes longer, often in the order of 10^{2} s [97]. Since the molecules lose some of the absorbed photons' energy via vibrational relaxation, the fluorescent light is often lower energy resulting in a higher wavelength. This shift in wavelength between absorbed and emitted light is called the Stokes shift and is a characteristic property for a molecule.

Based on previous research of Kolodner et al. and Haugen et al. [90–93], the two rare-earths: europium tris[3-(trifluoromethylhydroxymethylene)-(+)-camphorate] (EuTFC) and europium(III) thenoyltrifluoroacetonate (EuTTA) were selected as candidates for carrying out measurements at 77 K for investigating the quench in HTS tapes. Both compounds exhibit fluorescence, where the sample has to be continuously excited to produce an emission. The exact absorption, emission and temperature dependence of the fluorophores are discussed in Section 3.2.2.

3.2.2 Fluorescent spectroscopy and temperature dependence

The literature mentions optimal excitation light sources for the two fluorescent compounds researched here as well as their emission spectrum, however, it was necessary to confirm these for selecting optimal experimental devices. A spectral

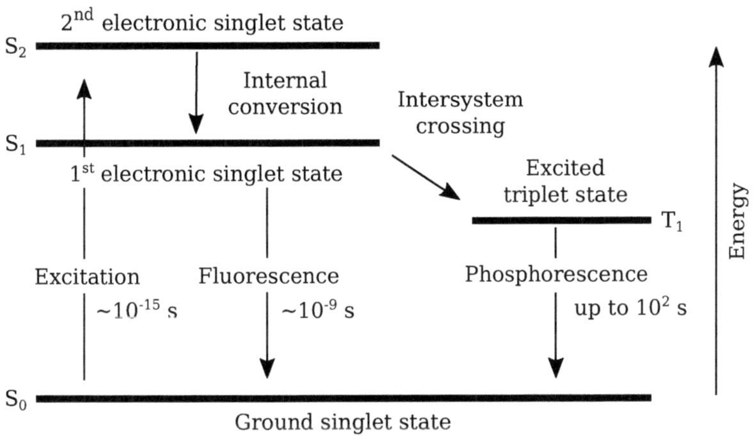

Figure 3.4: Simplified diagram showing the mechanics of excitation, fluorescence and phosphorescence including the representative duration of these effects. The figure only shows internal conversion between electronic states and intersystem crossing from singlet to triplet state, vibrational relaxation within the electronic levels is not shown.

analysis was done initially on the europium solution at room temperature for both its emission and absorbance. The results are shown in Figure 3.5a, using arbitrary units for the absorbance of the fluorophore. This is defined by the Beer-Lambert law

$$\alpha = \log(\frac{I_0}{I_1}), \tag{3.4}$$

where I_0 is the intensity of the incoming light and I_1 is the intensity of the light measured after passing through the sample. Hence, an absorbance of 1 equals to 90 %, whereas an absorbance of 3 to an absorption of 99.9 % of the incident light by the sample. It can be seen that over 90 % of the light between wavelengths of 360 nm and 400 nm is absorbed with the absorbance peaking at 376 nm. The noisy part below 380 nm in the EuTFC sample is likely caused by the solvent.

Figure 3.5b then shows the emission spectrum of the two selected fluorophore solutions at room temperature. The light intensity is here as well in arbitrary units, however it is a linear scale and simply represents a digitalization and has no physical units. The sharp emission peaks of both rare-earths are visible around 612 nm. The relative levels of fluorescence however show that EuTTA is approximately twice as bright at room temperature.

Based on this information as well as availability, compatibility and price of different components the light source as well as an optical filter were chosen. This latter is required to block any light that is not strictly caused by the fluorescence and hence improves the measurement accuracy and signal-to-noise ratio.

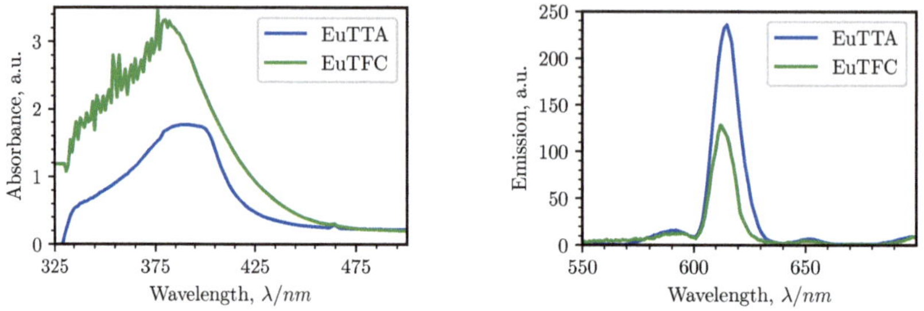

(a) Absorbance

(b) Emission

Figure 3.5: Measured (a) absorbance and (b) emission spectra of EuTTA and EuTFC in a polymethyl methacrylate solution at room temperature [99]. (a) shows an absorption of over 90 % of the incoming light between 340 nm and 400 nm for both compounds. In (b) the principal emission peak at 612 nm is visible when using 380 nm excitation light.

3.2.3 UV excitation light

Measurements with fluorescent thermal imaging in the past commonly used either Hg-Xe lamps [92, 93] or Hg-arc lamps [90, 91, 100] in microscopy setups as the excitation source. The light of the lamp was filtered, only allowing the necessary UV component to pass through and directed towards the sample using an optical guide. Such optical guides, however, generally only allow a smaller area to be illuminated due to their narrow cross-section.

However, advances in lighting technology, such as light emitting diodes (LEDs), allow for a much more compact and focused illumination of larger surfaces. For this reason, as well as due to size and place limitations, instead of a Hg-Xe or Hg-arc lamp, UV LEDs were used in the experiment with a central wavelength of 365 nm. At this excitation wavelength over 90 % of the light is absorbed by the fluorophore, as shown in Figure 3.5.

The excitation light intensity is of paramount importance for fluorescent measurements, especially when the measurement speed is of importance. In static measurements the exposure time of the camera can be virtually arbitrarily chosen, furthermore, additional post-processing can be applied to improve the final image quality, such as adding subsequent exposures. Haugen et al. [92] have achieved 75 mK resolution using an exposure time of 2 s at 84 K and previously Kolodner et al. [90] have used 20 s exposures and then combined these into a total effective exposure of 320 s to improve temperature resolution to 8 mK.

Achieving such high resolution comes at the price of measuring speed, as illustrated by the several second long exposure times used by these research groups. When transient effects of the duration in the few tens of milliseconds are to be measured, the thermal resolution decreases unquestionably. As the recording speed increases, the recorded light intensity becomes more and more faint and hence severely limits the accuracy. This can, of course, be compensated by either increasing the thickness of the coating or by increasing the light intensity of excitation light source. The satisfactory excitation light intensity was found experimentally by a trial and error process. UV LEDs mounted on a star-shaped circuit board from the company LED ENGIN were selected due to their wavelength characteristics, easy handling and soldering as well as affordable price [101]. A total of 8 UV LEDs with a typical output of 800 mW/unit, resulting in an approximate combined radiant output of 6400 mW, were found adequate for measuring at speeds of around 2500 frames per second (fps). However, it must be noted that a great portion of the radiant flux is actually lost during the measurement as stray illumination, that is light not reaching the sample but the nearby objects. Furthermore, since the light intensity is inversely proportional to the square of the distance, the exact light intensity at the surface is difficult to tell. During experiments the light intensity and camera speed were adjusted such that at the selected recording speed the dynamic resolution of the high-speed camera was nearly saturated. This allowed using the camera's full digital resolution.

It was also found in initial experiments that UV light can propagate though lengths of tens of centimetres in liquid nitrogen without considerable attenuation, compared to attenuation in air. The issue of liquid nitrogen boiling at room temperature will be discusses in Section 3.2.8.

3.2.4 Emission wavelength filtering

Since the emission peak of the chosen europium compounds is at 612 nm, an optical filter was used to filter out light apart from this wavelength. For this, the nearest matching 610 nm bandpass filter was used with a 10 nm full width at half maximum (FWHM) value. On the one hand side this ensures that ambient light is even further reduced – improving the signal-to-noise ratio – and on the other hand it also filters out any reflected component of the UV light from the LEDs that would otherwise cause additional noise in the camera's sensor. Requiring such precise filtering in industrial high-speed cameras is not exactly "state of the art" and therefore such filters usually have different mounting options and are not readily supported by the camera's f-mount. While in traditional photography a filter can be easily mounted in front of the camera's lens, this gives raise to central wavelength (CWL) shifts in the filtering for light beams with shallow angles, such that would happen at the edges of recorded images

$$\lambda_\theta = \lambda_0 \left(1 - \frac{n}{n_*} \sin^2\theta\right)^{\frac{1}{2}}, \tag{3.5}$$

where λ_θ is the CWL at a given angle θ ($\theta = 0$ being 90°, perpendicular to the filter), n the refractive index of the surrounding medium and n_* the refractive index of the filter itself.

Upon consultation with filter manufacturers about the problem and due to the unavailability of suitable lens-mounted filters it was found best that the filter be mounted in between the lens and the camera's mount using a custom 3D-printed frame. This allowed the light first to be collected by the lens and then to pass through the filter and onto the camera's CMOS sensor. A picture of the lens with the inserted filter is shown in the Appendix, Figure A.1.

3.2.5 Image acquisition using the high-speed camera

For image acquisition a Memrecam HX3 high-speed camera was used. The most important characteristics with respect to the high-speed fluorescent thermal imaging are listed in Table 3.1.

Since the developed fluorescent thermal imaging method only requires the imaging of light intensities and not the colours themselves, using a monochrome camera is sufficient. The factors that impact the final result are rather the data acquisition speed, the resolution at various speeds as well as the digital resolution. The Memrecam HX3 high-speed camera used for the measurements has a digital resolution of 12 bits, meaning that the analogue light signal can be digitized into $2^{12} = 4096$ units. This means a practical scale of 0 to 4095, where 0 represents absolute darkness (no light) and 4095 represents a fully exposed pixel in the camera's CMOS sensor.

The camera's speed can be increased by reducing the recording resolution and vice versa as shown in Table 3.1. So a resolution of 1920 × 1080 can be recorded at up to

Table 3.1: Important parameters of the Memrecam HX3 high-speed camera.

Property	Unit	Value
Resolution	pixel	1920 × 1080 (at 4670 fps)
		1280 × 720 (at 10 230 fps)
Digital resolution	bits	12
Sensor	-	CMOS
Colour	-	Mono
Storage	GB	16
Recording length	min	~12 (at 50 fps, 500 × 500)

4670 fps, yet by lowering the resolution to 1280 × 720 the speed can be increased up to 10 230 fps. From experience, using fluorescent light intensities in measurements a recording speed of 2000 fps to 7000 fps was adequate for recording quench propagation and provided in most cases almost saturated pixels at liquid nitrogen temperatures. It is important to mention that while the recording can be made in very high resolution, an object such as an HTS tape may only occupy a fraction of the total image. Commonly for example a 15 cm × 1.2 cm HTS tape had a resolution of approximately 450 × 150 pixels, whereas in the case of the measured pancake coils (with a recorded surface area close to 100 cm^2) approximately 500 × 500 pixels.

Due to the high recording speed the camera needs to first store the recording in a fast, internal memory, which is then available for download after the end of the recording. While the post-processing is not particularly computationally intensive, the saving and transferring of the data make the analysis process relatively slow. The measurements – especially on coils – also generate tens of GB of data (time series of images) per experiment that need manual inspection in each case, adding to the time burden of data evaluation. However, the method could in principle be turned into a true real-time quench detection method by using a standard camera or localized spectrometers and much slower recording speeds, e.g. 20 fps, allowing to post-process the recorded images directly. A balance between temporal resolution and the detectable minimum temperature change would allow setting a finer temperature margin for an emergency stop in the case of a localised quench.

3.2.6 Fluorescent film coating method

In order to measure the temperature of an object, for example an HTS tape, its surface has to be coated with the fluorescent material. Ideally the coating has to be thin enough to not alter the thermal behaviour of the object to be observed, but at the same time sufficiently thick to produce a strong enough emission needed for the measurements.

While many coating methods exist for creating thin layers, coating larger surfaces uniformly with the fluorescent material proved challenging. Based on previous works of Barton and Haugen et al. [93, 102] droplet deposition was applied that yielded satisfactory results in terms of coating thickness and stability.

As a first step the fluorescent compound, EuTFC or EuTTA, is mixed with polymethyl methacrylate (PMMA) and acetone in a ratio of 1.3 wt%, 1.7 wt% and 97 wt%, respectively. This results in a high doping of a fluorescent material in a polymer matrix, highly diluted in a solvent. The high doping ensures that the fluorescence is strong while the polymer matrix is responsible for the mechanical stability of the coating. Since the final solution is almost purely acetone with a yellow taint from the europium compound, it can be brought onto the surface of an HTS tape by droplet deposition. Upon placing a droplet of the solution to the centre of the surface to be covered, the acetone spreads out and evaporates rapidly, leaving behind the europium and PMMA over the surface. The coating is followed by a heat treatment, where the tapes are heat treated at 175 °C for 30 min [93]. The heating and cooling profiles were arbitrarily chosen to have a 175 °C h^{-1} rise time, held for 30 min and then cooled down naturally to room temperature. The temperature calibration was then based on this specific heat treatment process. The final result is illustrated in Figure 3.6.

The quality of the coating varied naturally from experiment to experiment, however its effects are not as severe as expected. Although the light intensity emitted by the coating strongly depends on its thickness, these can be filtered and effectively cancelled out during the post-processing as described in Section 3.2.8. In an initial experiment a steel tape was coated with EuTTA, where one side of the target was masked with a piece of tape to make a sharp line in the coating. The resulting microscopy image is shown in Figure 3.7, where the red rectangles indicate the approximate positions where atomic force microscopy was used to determine the thickness of the coating. It is visible that the coating is not uniform over the left hand side of the dummy and that the variations in thickness along the transition zone at the centre are in the order of a few micro metres.

3.2.7 Temperature calibration

In order to extract temperatures from later measurements it is of paramount importance to obtain an accurate temperature calibration of the fluorescent compound to be used. For this purpose a sample coated with the rare-earth has to be accurately temperature controlled, excited with a suitable light source and its emission has to be recorded accurately.

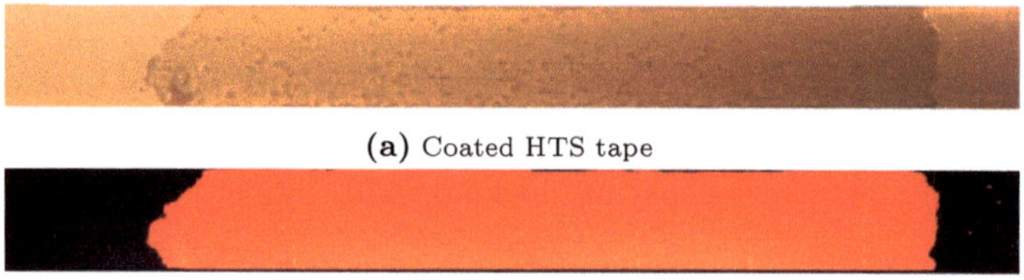

(a) Coated HTS tape

(b) Coated HTS tape in UV light

Figure 3.6: (a) shows an HTS tape coated with EuTTA, where the thin coating is vaguely visible over the centre of the tape. (b) shows the same tape under UV light excitation as visible for the human eye.

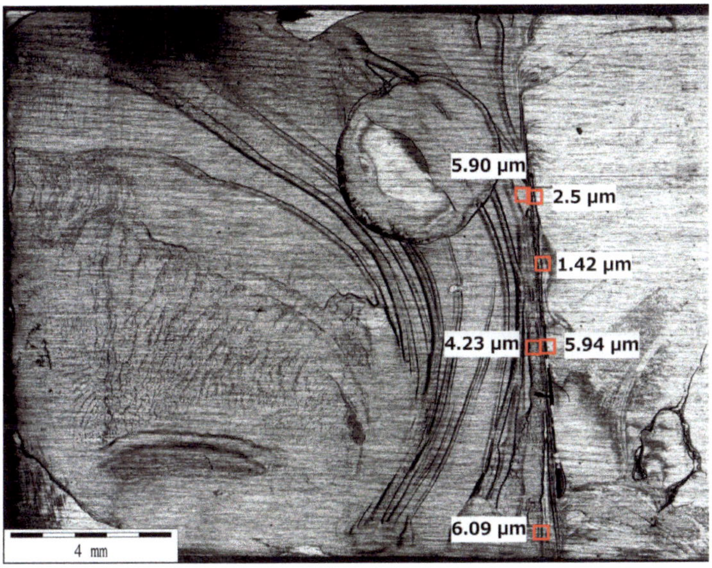

Figure 3.7: Microscopy image of a steel tape's surface coated until half-way (left) with EuTTA. The red rectangles show areas where atomic force microscopy was used to approximate changes in thickness [103].

Such an experiment was performed at KIT, at the Institute of Microstructure Technology (IMT) [99, 104]. A glass substrate was coated with EuTFC and cooled down to <4 K in an Oxford OptistatDry – TLEX closed cycle cryostat with two optical windows. A frequency doubled Ti:sapphire laser with an excitation wavelength of 365 nm excited the sample through the front optical window. Upon excitation, an AvaSpec – ULS3648 high-resolution spectrometer was capturing the fluorescent light through the other optical window (Figure 3.8). To keep consistent with later measurements, the same 610 nm filter was used in this experiment as in all future thermal imaging measurements. The cryostat's temperature was controlled by a LabVIEW program, where the continuous cooling was balanced by heaters to achieve a temperature above 4 K. Consequently, during the calibration measurement the sample was cooled first to 4 K and the light intensity was recorded. The temperature was increased to 10 K, during which the excitation laser's beam was optically blocked such that it no longer excited the sample. This was deemed important to cancel any possible effects of photobleaching – a process where due to prolonged excitation the fluorophore's molecules are changed and the fluorescence becomes progressively fainter – that would otherwise invalidate the measurement results. The sample was left for 5 min to reach thermal equilibrium at the new temperature, the light intensity was measured and the process was repeated in 10 K steps afterwards. With EuTFC after approximately 260 K the fluorescence became barely detectable and hence this temperature was chosen as the cut-off temperature for future fluorescent thermal measurements. During post-processing the results, the integral light intensity of the 610 nm ±10 nm was chosen as the final value for the visualization and calibration constant determination.

The measurement results are shown in Figure 3.9 where the light intensity is in arbitrary units as recorded by the spectrometer. Over the plotted range of 70 K to 260 K the fluorescent light intensity is decreasing almost linearly. The measurement of the EuTTA sample in the same experiment did not yield any usable relation between fluorescent light intensity and temperature, hence all measurements use the EuTFC compound. Also, even though that in the initial fluorescent spectroscopy at room temperature the EuTFC sample produced some fluorescence (Figure 3.5b), it was found that in practical measurement this light intensity is so low that only accurate spectrometers can detect it.

Since the present research focused on temperatures around 77 K, this region was investigated in detail during the calibration process. However, the method would work equally well at lower temperatures and using the full calibration curve until 4 K would allow calculating temperatures at this level as well.

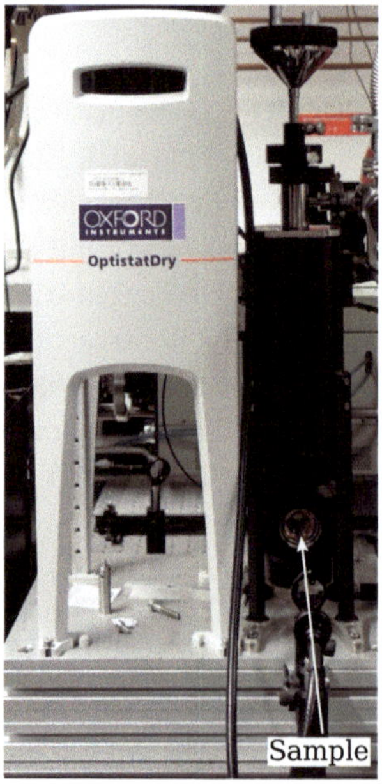

Figure 3.8: Optistat Dry cryostat used for calibrating the temperature dependant fluorescence in the EuTFC and EuTTA samples from 4 K to 300 K.

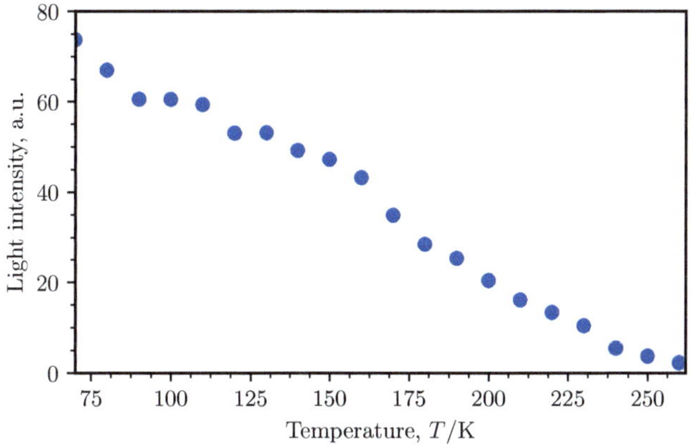

Figure 3.9: Light intensity (in arbitrary units) of EuTFC embedded in PMMA, baked at 175 °C for 30 min. The range of 4 K to 260 K was investigated in steps of 10 K [104]. Here only the range between 70 K and 260 K is shown, which is interesting for HTS coated conductors.

3.2.8 Extracting temperatures and post-processing

As previously mentioned, the method uses the light intensity change over a range of temperatures emanating from the fluorescent coating. Recording these changes allows calculating the temperatures, based on the previously determined calibration curve. The difficulty arises from the fact that the calibration measurement – for accuracy reasons – is usually done with a high precision spectrometer while the imaging itself is done with a high-speed camera. This introduces the problem that since the sensors are different, the digitalisation of the signals is also different and hence such arbitrary units cannot be compared directly. This makes a direct interpolation of measurement results over the calibration range impossible.

A solution to this problem was introduced by L. D. Barton [102] by using the relative light intensity changes instead of the absolute recorded values. Let $S(x,y)$ represent the light intensity at a point in a 2D image. This value depends on the: excitation intensity $I(x,y)$, the optical collection efficiency $\eta(x,y)$, the sample reflectivity $r(x,y)$, and the quantum efficiency $Q(T(x,y))$. The relation can be described as

$$S(x,y) = I(x,y) \cdot \eta(x,y) \cdot r(x,y) \cdot Q(T(x,y))\,. \tag{3.6}$$

In order to remove all optical artefacts, one can create a relative scale instead of the absolute measurement and divide the hot (biased) image, $S_{\mathrm{H}}(x,y)$, by the cold (reference) image, $S_{\mathrm{C}}(x,y)$, to get the ratio $S_{\mathrm{R}}(x,y)$,

$$\begin{aligned} S_{\mathrm{R}}(x,y) &= \frac{S_{\mathrm{H}}(x,y)}{S_{\mathrm{C}}(x,y)} \\ &= \frac{I(x,y)\cdot\eta(x,y)\cdot r(x,y)\cdot Q(T_{\mathrm{H}}(x,y))}{I(x,y)\cdot\eta(x,y)\cdot r(x,y)\cdot Q(T_{\mathrm{C}}(x,y))} \\ &= \frac{Q(T_{\mathrm{H}}(x,y))}{Q(T_{\mathrm{C}}(x,y))}. \end{aligned} \tag{3.7}$$

Since only the quantum efficiency of the fluorophore depends on the temperature, all other spatial optical artefacts cancel out. Hence, the local variations, in for example optical collection efficiency or excitation light intensity, are no longer a concern. This allows the calibration to be done on a different measurement device as the actual measurement. However, for all measurements a well-known reference temperature is required at the beginning that can serve as the basis for the normalisation.

Based on this principle first of all the calibration curve has to be normalized around the same temperature where future measurements will take place. In the case of objects immersed in liquid nitrogen this temperature can be taken as 77 K. The normalized calibration curve is shown in Figure 3.10. The temperature dependence was determined as

$$T(x,y) = \frac{S_{\mathrm{R}}(x,y) - 1.4250}{-0.005098}. \tag{3.8}$$

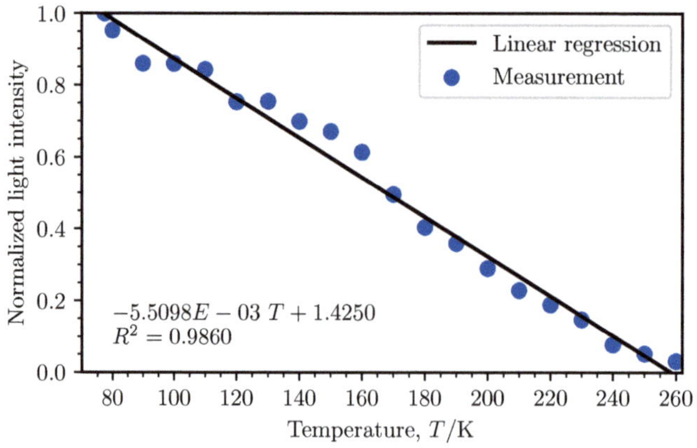

Figure 3.10: Light intensity of EuTFC embedded in PMMA, heat treated at 175 °C for 30 min and normalized around 77 K. The R^2 value shows a satisfactory fit to a linear approximation.

The same normalization is then to be carried out on the recordings as well, this process is shown in Figure 3.11. Firstly a reference frame is created at the beginning of the video, Figure 3.11a, which is usually the average of the first 1 to 10 frames. Averaging on a pixel-by-pixel basis helps average out the noise in the camera's sensor and effectively allows using a longer exposure-equivalent image quality for the reference. This reference frame is then used to normalize the rest of the frames in the video, also on a pixel-by-pixel basis. The actual frame, Figure 3.11b, is divided by the reference frame, where the tape is known to be at constant temperature (here 77 K since it is immersed in a liquid nitrogen bath, but the the tape could be held at any other reference temperature). As the light intensity of the fluorescence decreases with increasing temperature, the reference frame is always brighter than, or equal to, any of the later frames. The result of the normalization is shown in Figure 3.11c, where the scale of 0 to 1 simply depicts the relative light intensity change with respect to the reference frame. Here a value of 1 (white) means that no change has happened, whereas a value of 0 (black) would mean that the fluorescent light at the given coordinates was fully extinguished. The final step is the extraction of thermal data from the normalized frame, by interpolating the values on the previously normalized calibration curve as shown in Figure 3.10. The resulting scale in Figure 3.11d was therefore converted from the 0 to 1 range to the applicable light emission range of the EuTFC coating, being 77 K to 260 K in this case. Note that the colour of the plot is merely a representation of values and aids visualization.

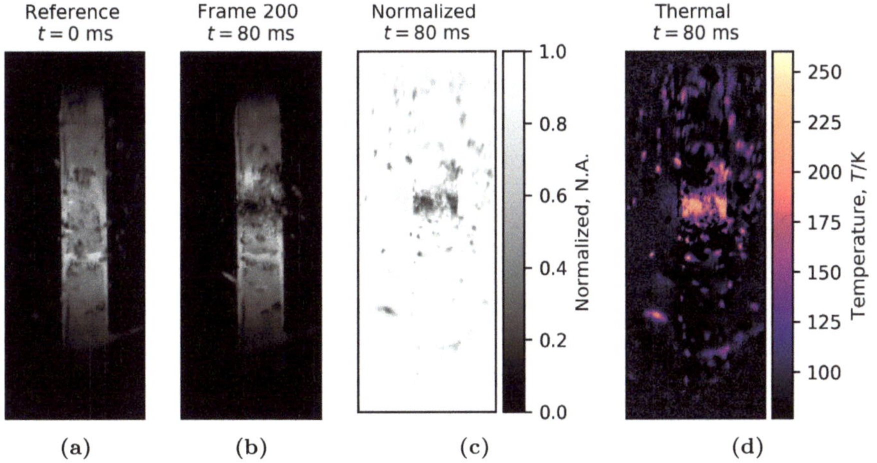

Figure 3.11: Post-processing steps illustrated on an EuTFC-coated HTS sample measurement. (a) shows a reference image. (b) shows frame 200 of the recording at 80 ms where some heating takes place at the centre of the sample. (c) shows frame 200 normalized with the reference frame. (d) shows the extracted thermal data from (c) using the calibration curve shown in Figure 3.10.

As shown previously, a decrease in light intensity means that the local temperature increased. Hence the expectation is that an object at a steady, known temperature will emit a certain light intensity and upon some local heat input the light intensity would decrease. This can be translated simply in bright = cold and dark = warm with respect to the recorded (as well as to the normalized) images. Figure 3.12 shows the effects of post-processing that have to be taken into consideration. The upper row, Figures 3.12a–3.12d, show excerpts of a recording at times 0 ms, 20 ms, 40 ms and 60 ms with the scale set to a digital resolution of 12 bits equalling 4096 brightness levels. The red lines and the arrow on the first image are a visual aid and show the position and movement of a selected nitrogen gas bubble. It is visible that as time passes, the bubble changes position and rises upwards. The same is shown in the lower row, Figures 3.12e–3.12h. These are the normalized frames, the quotient of the actual frame and the zero frame, and hence can only take a value between 0 and 1.

The two important things to observe are related to the fact that a transient effect is being recorded where movement takes place. In a stationary measurement the following problems would not be present. Both issues are related to the relative movement of the selected bubble and its appearance in different position on the reference frame and subsequent frames. Since light is refracted at the surface of a gas bubble, depending on its position it can either cause a brighter or a darker spot in the image.

The first observation is that while the gas bubble highlighted by the white arrow clearly moves in Figures 3.12a–3.12d and causes a brighter spot, in Figures 3.12e–3.12f it is stationary, as highlighted by the black arrow. The bubble's position in the reference, frame 0, has a higher brightness value than the same position in any subsequent image (since the bubble has moved away). This would normally result in a division where the pixel value is

$$\frac{S(x,y)|_t}{S(x,y)|_{t+\Delta t}} > 1 \tag{3.9}$$

at the new position of the bubble and hence a second bubble should be visible. This is shown with a red arrow in Figure 3.12h where the bubble should be. However, as discussed before, the zero frame at a steady cold temperature has to be always brighter than any later image (assuming no cooling below the reference temperature). Hence the only explanation for a brighter spot in a later frame at the same position could be either some form of noise from the measurement device or some disturbance of the system. Since noise would not cause such huge changes, the cause can be determined as a disturbance, the bubble in this case. This problem is fairly straightforward to mitigate, an "if-else" statement in the program can check for pixels in the video that got brighter and simply replace them with a value of 1 in the normalized frame.

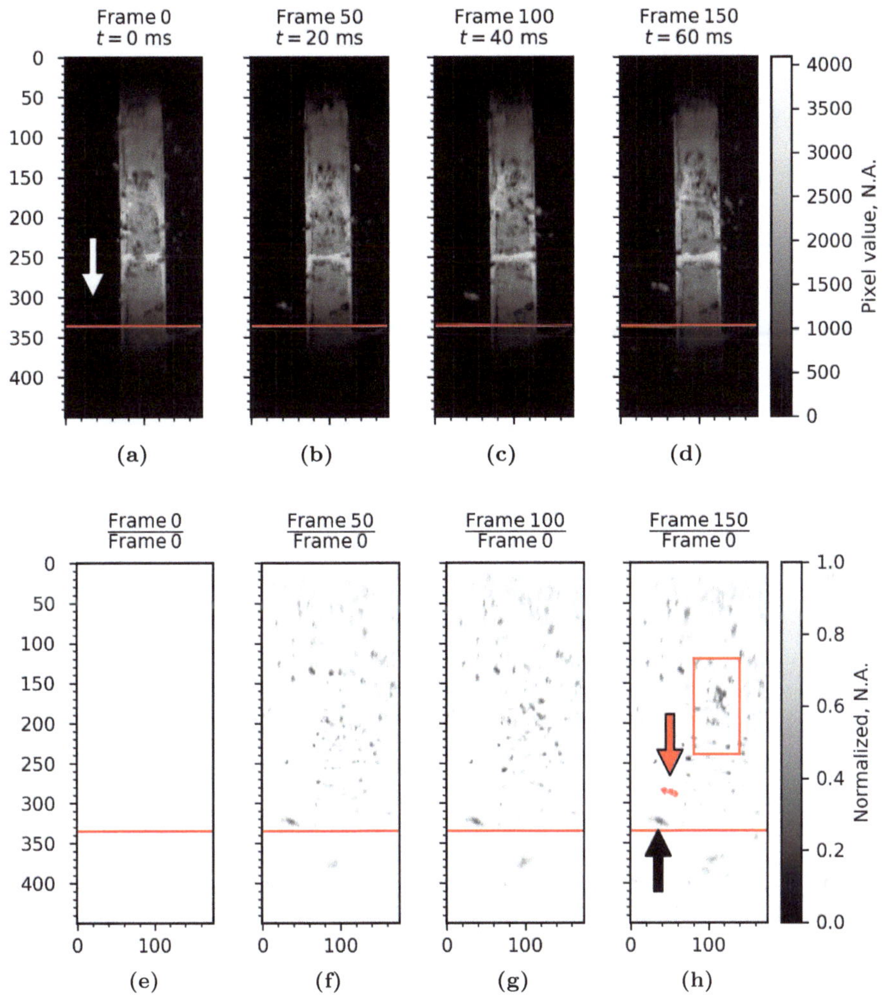

Figure 3.12: Excerpts of a recording on an EuTFC-coated HTS tape placed in a liquid nitrogen bath in a vertical position. The effects of normalization are shown in this figure. The upper four images, (a)-(d) show single frames as recorded by the camera. The lower four images, (e)-(h), show the software post-processed, normalized images, where the same frame is divided by frame 0. The highlights show areas of interests and artefacts after post-processing.

The second observation is that the normalized images retain a "burned-in" position of gas bubbles as in the reference frame. Since the divisor image has artefacts on it (a bubble on this zero frame) where the light intensity is higher than in later frames, this position will always show a value considerably smaller than 1, as shown by the black arrow in Figure 3.12h. As a consequence, the quality (distortion due to gas bubbles) on the reference frame at the beginning of the recording has a large impact on the quality of the entire recording.

The discussed gas bubble causes a brightly moving spot in this specific case, however, bubbles can also cause a lowered light intensity. This is visible for example in Figure 3.12h, highlighted by the red rectangle. Compared to this, the same area in Figure 3.12f and Figure 3.12g has fewer dark spots. The reason in this case is that in frame 150 (3.12d) a larger number of small gas bubbles are present in this region that cause partial darkening. In this case the situation is the exact opposite and the normalization at this position would result in a division

$$\frac{S(x,y)|_t}{S(x,y)|_{t+\Delta t}} \ll 1, \tag{3.10}$$

meaning a dark point as visible in Figure 3.12h. This error is practically impossible to correct since if a heating takes place simultaneously, it is not know how much of the light intensity decrease is due to the heating and how much due to the refraction due to the gas bubble.

The final observation to be made is that as shown in eq. 3.7, stationary optical differences due to light intensity or coating thickness are cancelled out. This is visible in the upper row in Figure 3.12, where a brighter, horizontal strip is visible around the centre of the tape. This is most certainly caused by a thicker layer of coating, however this effect is no longer visible in the normalized images in the second row.

3.3 Temperature resolution and accuracy

Even though the high-speed fluorescent thermal imaging method readily provides qualitative information about the surface temperature of HTS tapes during a current pulse, evaluating the accuracy and thermal resolution is important.

Due to the acquisition speed alone judging the method's temperature resolution and accuracy proved challenging. While the resolution can be calculated according to eq. 3.12 for any recording depending on the maximum brightness in the given situation, accuracy does not let itself be defined so trivially. The accuracy depends strongly both on the calibration curve as well as on the individual measurement.

3.3.1 Noise

When a recording is made of a completely dark environment, by closing the camera shutter, the CMOS sensor does not record a value of true 0, but some higher value. This value is not constant over the sensor, but varies slightly over time and over position as well. Some areas of the sensor might record higher variations, which is referred to as fixed pattern noise (FPN), and can be cancelled using post-processing (this was done by the camera automatically upon exporting a recording). The remainder of the noise is simply dark noise of the sensor and, being random, cannot be removed from the recording or individual images. To get an idea of the noise in the camera's sensor, a short recording was made at two different frame rates, both with and without the low-light mode (called "GXC" [1]), with the camera's shutter closed. This blocks all light from reaching the sensor and hence the recorded values are simply the noise. An example is shown in Figure 3.13, with noise values similar to that in actual recordings, and an actual measurement in Figure 3.14 shows the maximum relative error

$$\varepsilon_{\text{rel, max}} = \max(S(x,y)|_{t+1} - S(x,y)|_t) \tag{3.11}$$

appearing in a video over time.

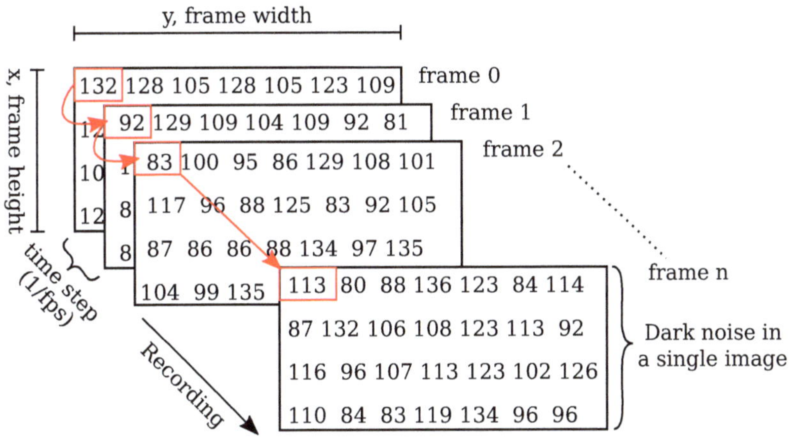

Figure 3.13: Representation of a grayscale recording with a camera showing random values, such as dark noise in the sensor of the high-sped camera. The red rectangles and arrows highlight the relative change in recorded light intensity of a selected pixel in the image.

[1] This setting was designed specifically for measurements in low light conditions and it effectively increases the light sensitivity by four fold in the recordings, whereas the noise level only increases two-fold.

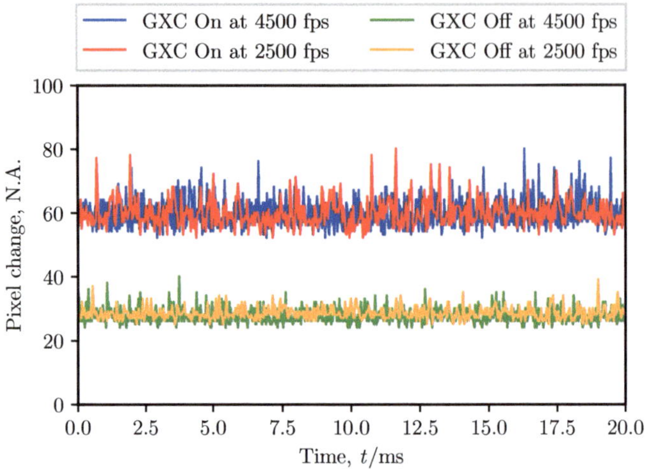

Figure 3.14: Time evolution of dark noise in the camera with closed shutter at two distinct recording rates with and without GXC. The levels represent the single largest change of recorded light intensity of any pixel with respect to its prior state.

Figure 3.14 shows that while the recording speed does not influence the noise level, enabling the GXC mode effectively doubles it. However, even though the recording speed does not influence the noise directly, since it is related to the recorded light intensity, it effectively changes the maximum recorded value and thereby the relative error as well.

3.3.2 Thermal resolution

The maximum thermal resolution of the method is determined by the digitalization of the measured light intensity and the temperature range of the measurement. As previously discussed in Section 3.2.5, the high-speed camera has a 12 bit digital resolution, meaning that a maximum of $2^{12} = 4096$ light intensity levels can be recorded in the range $[0, 4095]$. The camera's special "GXC" mode also has the effect that although the camera records in 12-bit mode, the actual light intensity values saved in the final 16-bit TIFF file can have light intensities above the 12-bit limit of 4095, due to the post-processing of the camera itself. Assuming a maximum recorded value (of a selected pixel) of 4095 and a minimum value of 0, the sensitivity can be computed as

$$\text{Sensitivity} = \frac{T_{\max} - T_{\min}}{4096} = \frac{260\,\text{K} - 77\,\text{K}}{4096} = 0.0447\,\text{K}. \tag{3.12}$$

On the one hand a recorded value of 0 is practically impossible due to various noise sources in both the measurement as well as in the camera's electronics. On the other hand a value of 4095 is also just an approximation for the maximum light intensity as for example due to the GXC mode of the camera intensities above this level can be realized as well. Since this maximum recorded light intensity is a function of sample-to-camera distance, UV excitation light intensity, average coating thickness, position, sample temperature and recording speed, so is the sensitivity in every measurement. Therefore it can be more useful to define the accuracy as a function of noise and usable digital resolution as in Figure 3.15, based on eq. 3.12. Here the red and green curves show the theoretical sensitivity as well as the realistic sensitivity over a 2×2 pixel area. The blue and orange curves show the single pixel sensitivity when using the camera without and with the low-light GXC setting, respectively. Since the observed temperature range is only 183 K, disregarding noise, even low digital resolutions would result in reasonable accuracy using the given camera. Accounting for noise, however, can reduce the accuracy locally to several kelvins. Nevertheless, even though the relative error of a single pixel may be in the order of 2 K, the accuracy can be improved by averaging over a small area. Averaging for example over a 2×2 pixel area already reduces the noise by a factor of ~2.7 compared to the single pixel noise of 30.

It is important to note, however, that this sensitivity and relative error level is specific to the high-speed camera used in this work.

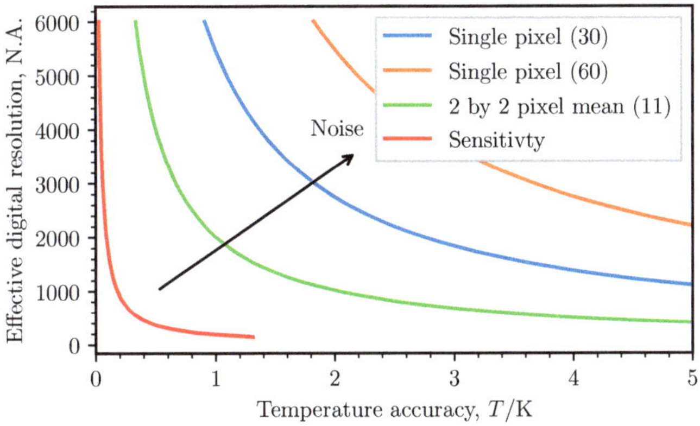

Figure 3.15: Temperature measurement accuracy as a function of effective recording digital resolution. The red curve shows the theoretical sensitivity, while the green, blue and orange lines show a 2×2 pixel mean, and a relative noise of 30 and 60, respectively. The black arrow shows the effect of increased noise in the recordings.

4 Quench propagation and analysis in HTS tapes

The application of the previously introduced high-speed fluorescent thermal imaging method is discussed in this chapter from a technical point of view. In particular, the measurement assembly, data acquisition and the measurement protocol are introduced. This is then followed by measurement results of various superconductor tapes that are analysed in detail, highlighting aspects such as the normal zone propagation velocity, temperatures during a quench as well as the effect of defects or special architectures in coated conductors. The purpose of the introduced method is not merely to measure the NZPV in HTS tapes, but also to provide a qualitative investigation of the quench and its propagation at a fine temporal resolution. Hence this chapter focuses on previously not investigated or not well-defined and understood aspects of quench in HTS.

4.1 Experiment description

Figure 4.1 shows a detailed cross-sectional view of the measurement setup used for short sample measurement. The large, outer enclosure – originally a cryogenic bath itself – was used for the sole purpose of blocking out ambient light. The actual measurement cryostat, a double-walled, glass cryostat was placed inside the larger assembly and housed the sample holder. The high-speed camera was facing downwards in another, cylindrical cryostat (part of the larger cryogenic bath assembly), and viewed the sample holder and the HTS tape through a mirror. The excitation UV light was positioned outside the cryostat, close to the glass interface and excited the sample at an oblique angle from a short distance, typically ~10 cm. The glass cryostat was filled with liquid nitrogen and a short, dc current pulse was applied across the sample. Then, at the same time, a high-speed recording was taken together with electrical measurements of the sample. By using a magnetic defect at the centre of the sample, the quench formation and development could be directly observed in the recording. Finally, with the help of post-processing the temperatures during a quench could be calculated and the heat propagation was visualized.

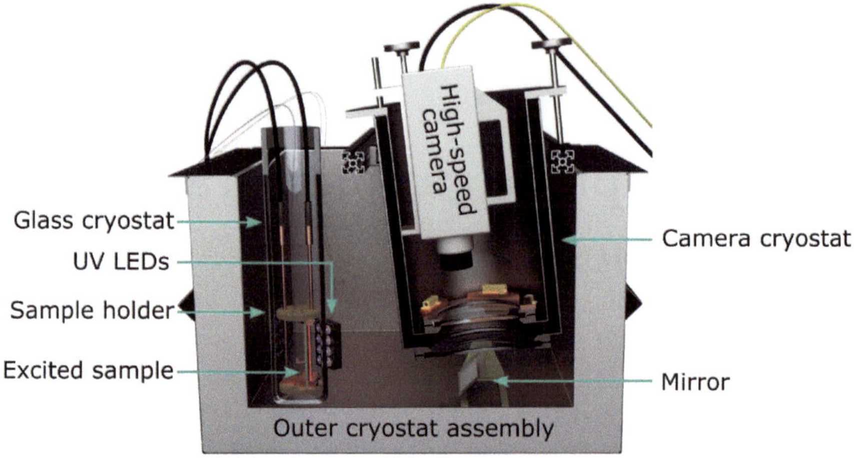

Figure 4.1: Cross-section view of the experimental assembly for HTS tape measurements. A larger cryostat bath is used as an outer enclosure, holding a double-walled glass cryostat, a submersible, cylindrical cryostat for the high-speed camera as well as the camera itself.

4.1.1 Experimental setup

The first requirement for the fluorescent thermal imaging method was a cryostat to cool down HTS samples to 77 K and simultaneously allow the transmission of both visible and UV light. In previous experiments using the high-speed camera [105, 106], the large volume liquid nitrogen bath cryostat was used with a submersible secondary cryostat hosting the high-speed camera (outer enclosure shown in Figure 4.1). This method, however, made it difficult to position the UV LEDs such that they would provide sufficient light intensity as required for recording speeds in the order of several thousand frames per second. In this configuration, first the UV light had to cross 40 cm to 50 cm of liquid nitrogen and then the emitted light the same distance. This caused large attenuation in the recorded light intensity and was hence deemed unfit for the purpose. Instead, a much smaller volume glass cryostat was designed and manufactured at KIT. This was a 60 cm long double-walled cryostat made with an inner diameter of 10 cm from borosilicate glass, which allows transmission of UV light. The space between the inner and outer wall was evacuated to form an insulation. The experimental layout shown in Figure 4.1 allowed placing the UV light in close proximity of the cryostat (and the HTS sample), which then made recordings at up to 9000 frames per second possible.

From the beginning, efforts were made to reduce the natural boiling of liquid nitrogen at atmospheric pressure, as to avoid optical artefacts in the recordings as introduced in Chapter 3. Unfortunately, however, this could not be accomplished to the full extent. The cryostat itself was not perfect and due to some impurities (either material defect, such as a scratch, or introduced during manufacturing) there was a nitrogen gas nucleation point at the bottom of the cryostat. This caused a "jet of gas bubbles" to rise from the bottom of the cryostat and pass in front of the HTS sample. The problem was (at least partially) resolved by placing a semicircle-shaped Teflon sheet at the bottom of the sample holder, just below the HTS sample, at a 45° angle as shown in Figure 4.2. This then deflected the majority of the nitrogen gas bubbles rising from below the sample holder to one side of the cryostat, away from the HTS tape sample at the centre. At the same time, the UV LEDs were placed on the opposite side, relative to where the "gas jet" was deflected to, such that the incoming UV excitation light would not get time-dependent scattering before exciting the sample surface.

With time the evacuated space between the cryostat walls partially "lost" the vacuum insulation and condensation on the outer glass surface became an increasing problem due to humidity in the ambient air. This problem was mitigated by applying a stream of room-temperature air at the surface of the glass cryostat to prevent the moisture from condensing. Boiling in the cryostat could not be avoided to full extent, which created distortions during the measurements, as it will be seen in this chapter. As introduced previously in Chapter 3, these distortions are time-dependent and hence cannot be cancelled out during post-processing. Consequently, the quality of the recordings varied between measurements.

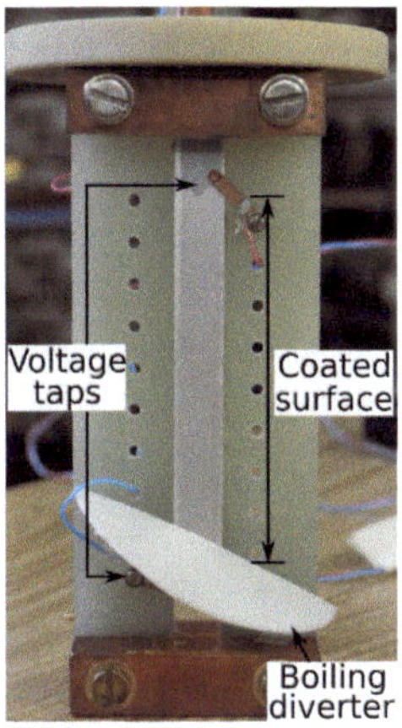

Figure 4.2: Close-up of the sample holder showing a 12 mm stabilizer-free sample placed at the centre with two voltage taps. The approximated length of the surface with the fluorescent coating as well as the "boiling diverter" are shown.

4.1.2 Sample preparation

The measured HTS samples were generally 150 mm long, from which the current leads occupied a total of 2 × 15mm, leaving a visible area of 120 mm. After the sample was coated with the fluorescent coating (as introduced in Chapter 3), it was placed in the sample holder, shown in Figure 4.2. This consisted of a glass-fibre reinforced plastic (G10) board with a copper current contact terminal at the top and bottom. A cylindrically-shaped G10 disk, perpendicular to the sample holder board, helped positioning the sample holder inside the cylindrical cryostat and at the same time prevented it from moving. The HTS tape sample was placed on the G10 board, which hosted a small hole with a 3 mm × 5 mm, cylindrical magnetic defect at its centre. The purpose of the magnet was to locally reduce the I_c of the coated conductor and ensure that the quench nucleation happened at a well-visible position. The top and bottom of the tape was clamped with the copper current contacts, using a thin sheet of indium foil in between the surfaces, which provided satisfactory contact characteristics at current pulses going above 800 A.

Most of the unobstructed length of the HTS was coated with the fluorescent layer, which due to the PMMA matrix is electrically insulating. This made it difficult to measure simultaneously the quench propagation from a thermal and electrical perspective, especially if one does not want to obstruct the view with voltage taps. For this reason, only a single pair of voltage taps was used, further apart, outside of the coated region. The purpose was not to measure the NZPV, but simply to get an electrical feedback about the quench in the tape and the developed electric field. For measuring the voltages, flexible pogo pins were used.

On special requests from the FASTGRID project [107] and the company Oxolutia, experiments were made with shorter sample lengths, however, such samples proved to be challenging to measure. In particular, problems were encountered with samples of 50 mm to 60 mm in length. In such cases, 2 × 15mm length of current contacts would have reduced the usable, visible area to almost nothing. To maximize the visible area, in some experiments the current leads were only contacted over a length of 5 mm to 10 mm to allow for the voltage taps and fluorescent measurement. Such a small contact area proved problematic and often caused a burnout of the tape – or serious I_c degradation – under the current leads. This was especially the case with tapes using artificially increased surface resistivity. It was found that for the measurements and for avoiding damage to the tapes, a minimum sample length of 100 mm is advisable.

4.1.3 Measurement procedure

The aim of the investigation was to see the development of the quench and NZPV in HTS tapes, preferably without destroying the samples. To have fine control over the time domain of the measurements and to avoid heating due to continuous currents, pulsed current measurements were used. Short – tens to hundreds of milliseconds – direct current pulses were applied to HTS tapes and simultaneously the voltage development across the tape was recorded, parallel to high-speed fluorescent thermal imaging. The shape of current pulses was similar to that shown in Figure 4.3. The pulse duration and amplitude were both varied based on the HTS tape's characteristics, most importantly the I_c and the tape architecture. To avoid accidentally damaging the samples, commonly the pulsed measurements started slightly below the I_c of the sample and in subsequent measurements the amplitude was gradually increased. When a quench was first observed, further tests were carried out by adjusting both the amplitude and the duration and see the effects.

It is the deposited energy Q in the tape over the duration of a pulse (not to be confused with the MQE) that causes heating and inevitably a quench, governed by

$$Q = \int_0^{t_p} P\mathrm{d}t = \int_0^{t_p} I \cdot V\mathrm{d}t. \tag{4.1}$$

Here P is the dissipated power and t_p is the duration of the current pulse. From experience, pulse durations close to I_c of more than 50 ms have a high probability of permanently damaging non-stabilized tapes. As a consequence, most conductors were commonly pulsed for approximately 10 ms. Tapes with metallic stabilization are inherently more stable and can bypass more current over the metallic shunt. In such tapes much longer pulses could be applied, on occasion up to 300 ms, without damaging the tape. Nevertheless, several hundred millisecond long pulses were already long

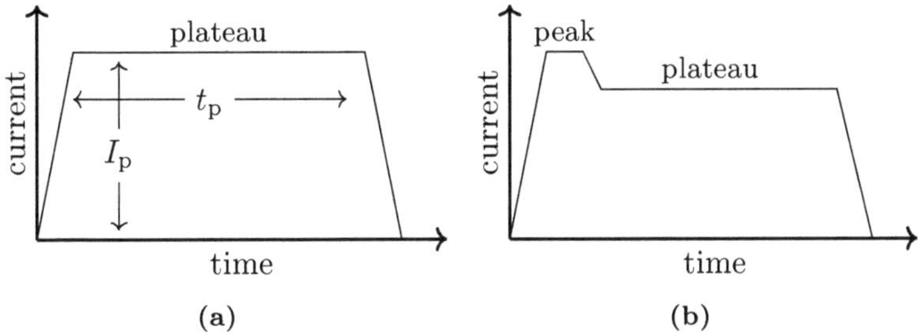

Figure 4.3: Direct current pulse shapes applied to HTS tapes. (a) shows a rectangular current pulse. (b) shows a rectangular current pulse, with an initial peak at a higher amplitude for a short duration.

enough for nitrogen gas bubbles to form on the heating copper current contacts at the bottom of the measurement probe. This caused nitrogen gas bubbles to rise upwards and severely distort the view making the acquisition of useful data near impossible. For these reasons, in most measurements – stabilized and non-stabilized tapes – shorter current pulses were preferred close to and above the I_c of the respective sample.

A particular pulse shape is shown in Figure 4.3b, used to quench tapes with an even higher stability. The initial, larger current amplitude would trigger a quench, and by rapidly dropping the current to a lower level, damaging the sample could be avoided.

The rising and falling times of the current pulses were adjusted empirically to be as fast as required. Determined from experimental results, the pulsed source can deliver currents up to 850 A within ~1.4 ms. For occasional higher currents a rising time of ~2 ms was used. Trying to request the peak current faster than these levels caused switching noise and occasional overshoot in the source. Dropping the current at the end of the pulse was possible within ~200 µs, however, generally the same rate was used as for the ramp up.

4.1.4 Data acquisition

Since the measurements were done using short current pulses, apart from high-speed recordings, fast electrical data acquisition was also required. Figure 4.4 shows a simplified diagram of the data acquisition procedure. For the acquistion of analogue signals an 18-bit NI USB 6289 data acquisition unit (DAQ) was used with a maximum available sampling rate of 625 kS s^{-1} together with a Dewetron amplifier using four DAQP-LV-B modules. The modules have an adjustable amplification factor of up to 500-fold, which allowed fine-tuning the measurement sensitivity versus range. The amplifier also served as a galvanic insulation between the measured voltage signals and the data acquisition unit, protecting the DAQ from potentially high voltage spikes. A LabVIEW software was used to pre-generate a finite duration, digital current waveform, which was then scaled to a voltage waveform and finally converted to an analogue output signal by the same DAQ. The analogue voltage waveform was fed into a Heizinger pulsed current supply with a maximum output current of 1650 A and a maximum voltage of ~10 V. The current supply's output followed the input voltage and hence the requested current pulse was obtained.

The start of the measurement was signalled by a transistor-to-transistor logic (TTL) SYNC signal, which was fed to the pulsed current source using a serial connector. Simultaneously, the SYNC signal was split off to trigger the high-speed camera and was also fed into the data acquisition using coaxial cables. By electrically recording

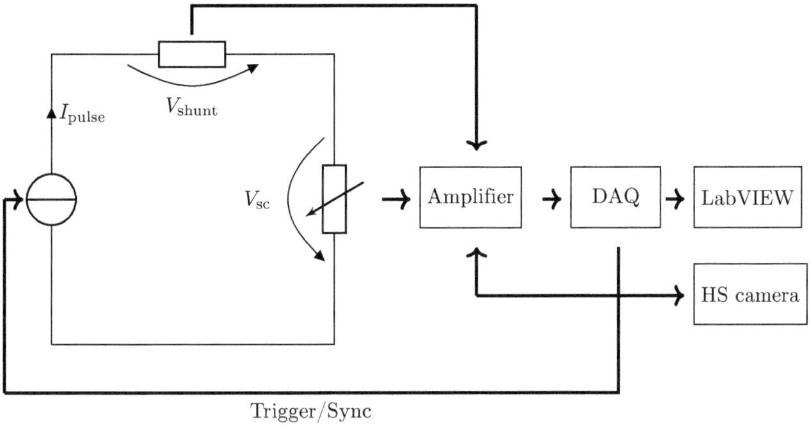

Figure 4.4: Data acquisition of various voltage signals in the pulsed quench measurements.

the SYNC signal, the beginning of the recording could be matched with the electrical measurements with the help of accurate time-stamping.

The DAQ recorded three channels: current through a shunt resistor inserted into the circuit, voltage drop in the superconductor measured across two voltage taps and the reference SYNC signal going to the high-speed camera. The data acquisition speed was varied based on the pulse duration, generally speeds were in the range of $10\,\mathrm{kS\,s^{-1}}$ to $30\,\mathrm{kS\,s^{-1}}$ per channel.

The UV LEDs were powered by an Aim TTi QL355P dc power supply with up to their rated current of 700 mA [101]. The LEDs were only turned on for the short duration of the measurement and were turned off directly after a pulse was applied.

4.2 Optical NZPV measurements

While in the state-of-the-art NZPV measurement method the arrival of electrical signals is detected at precise distances, as explained in Chapter 2, in the case of an optical measurement another criterion is necessary for determining the NZPV. This section focuses on discussing the applicability of the method to determine the NZPV from a thermal perspective, calculated purely from the recorded raw fluorescent or the post-processed thermal images. A more detailed discussion of quench formation and behaviour follows in Section 4.3.

In order to cross-check the normal zone propagation velocity, electrical measurements from the group at Polytechnique Montréal, Canada were reported for a range of HTS samples. This was then compared with fluorescent thermal imaging measurements at KIT on either the same sample or a sample from the same batch of conductor.

Unfortunately, in the different measurement setups and conditions, the tapes were not always quenched exactly in the same way. Even though all efforts were made to apply closely identical current pulses, both in amplitude and duration, differences still existed.

4.2.1 Measurement method

To determine the normal zone propagation velocity, first the recorded images' axes were converted into millimetre scale. This was done by comparing the number of pixels along the width of the tape and using the number of pixel per centimetre ratio to also plot the longitudinal axis. Using the new, metric scale, a thin grid overlay was plotted on top of the recording, using lines in 2 mm distances. Then, the temperature scale was limited to the range 77 K to 93 K, using the critical temperature of REBCO superconductors[1] as the upper boundary. Here the assumption was made that sections of the tape transited to normal conducting state when the critical temperature is exceeded, compared to an empirically selected critical electric field, as in the case of electrical NZPV measurements. Finally, the processed image sequence was evaluated in each case manually, where the "flight of the 93 K zone" was calculated at different time steps starting from the first sign of appearance of the quench at the magnetic defect.

Figure 4.5 illustrates the calculation of the NZPV using the thermal imaging method. The position of the magnetic defect is known on the recording and hence the centre line from where the normal zone propagates x_0 as well. On a subsequent image at time i the propagation of the 93 K zone can be observed upwards until point x^i_{up}. From this, the NZPV was calculated as

$$\mathrm{NZPV}^i_{\mathrm{up}} = \frac{\Delta x}{\Delta t} = \frac{x^{\mathrm{i}}_{\mathrm{up}} - x_0}{t_{\mathrm{i}} - t_0}, \tag{4.2}$$

where Δx and Δt represent the spatial and temporal change in the propagation, respectively. If a visible difference was observed in the travelled distance by the normal zone in the opposite direction (downwards in this case), the NZPV was calculated at a later time $i+1$ as

$$\mathrm{NZPV}^{i+1}_{\mathrm{down}} = \frac{\Delta x}{\Delta t} = \frac{x^{\mathrm{i+1}}_{\mathrm{down}} - x^{\mathrm{i}}_{\mathrm{down}}}{t_{\mathrm{i+1}} - t_{\mathrm{i}}}. \tag{4.3}$$

[1] Common values used for the T_c of HTS coated conductors is in the range of 90 K to 93 K [6, 48, 73, 108, 109].

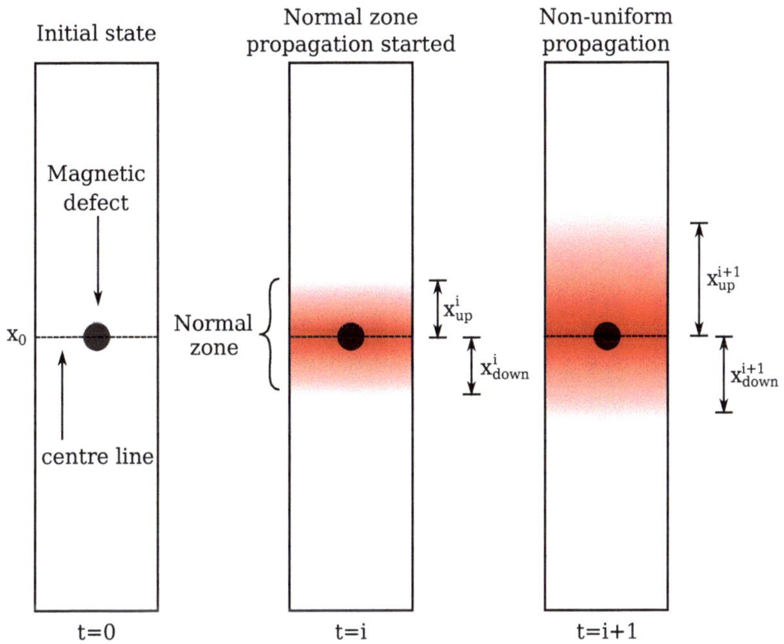

Figure 4.5: Normal zone propagation in HTS tapes showing the steady-state operation at 77 K as well as two time instances after the appearance of a normal zone. The red coloured area indicates the propagation of the 93 K front, used as a criterion for calculating the NZPV.

An actual example of a post-processed recording is shown in Figure 4.6, with the visual aid of the overlaid gridlines in 2 mm intervals. In this example the centre line, where the normal zone started developing, was at $x_0 = 42\,\mathrm{mm}$, as measured from the upper part of the tape. The normal zone development itself was first observed at approximately $t_0 = 7\,\mathrm{ms}$. Starting from this time and position, Figure 4.6a indicates that the normal zone propagated 4 mm up and 4 mm down from the nucleation point until the 12.57 ms mark. Then, an additional 2.57 ms later in Figure 4.6b (absolute time of 15.14 ms), the normal zone travelled 6 mm in both directions. By knowing both the elapsed time and distance travelled (with respect to t_0 and x_0, respectively), the NZPV was obtained using eq. 4.2 for the first time step

$$\mathrm{NZPV} = \frac{0.004\,\mathrm{m}}{0.012\,57\,\mathrm{s} - 0.007\,\mathrm{s}} = 0.72\,\mathrm{m\,s^{-1}} \tag{4.4}$$

and eq. 4.3 for the second time step

$$\mathrm{NZPV} = \frac{0.002\,\mathrm{m}}{0.015\,14\,\mathrm{s} - 0.012\,57\,\mathrm{s}} = 0.78\,\mathrm{m\,s^{-1}}. \tag{4.5}$$

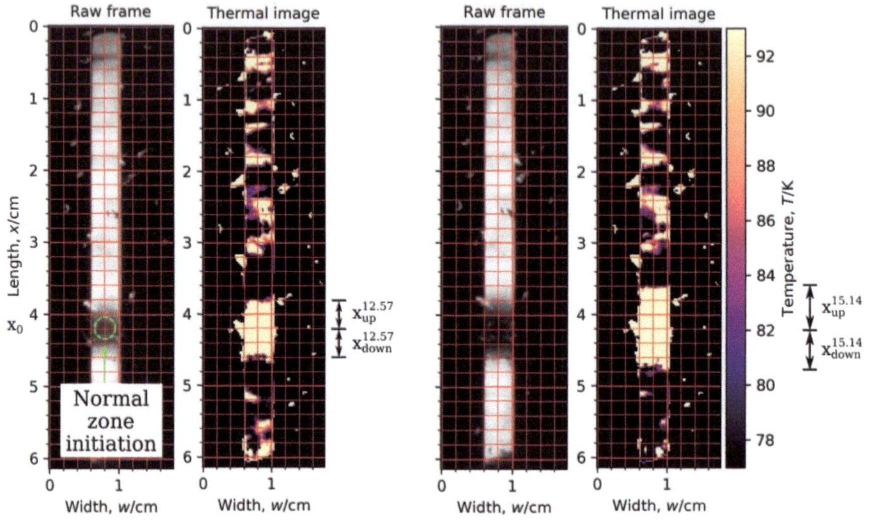

(a) t =12.57 ms

(b) t =15.14 ms

Figure 4.6: Example of a recording post-processed for calculating the normal zone propagation velocity. (a) shows the normal zone at time 12.57 ms and (b) at 15.14 ms. Both show the raw frame (left) and a thermal image (right), with the scale boundaries between 77 K and 93 K. The red grid lines are 2 mm apart and the arrows indicate the nucleation point (caused by the magnetic defect) and the distance travelled by the quench in the up- and downwards direction.

Images processed in this manner are difficult to use for other purposes and hence this post-processing was solely used as an intermediate step for calculating the thermal NZPV. For general visualization purposes and observing the behaviour of the quench the grid lines were not plotted.

It must be noted that – especially in the case of some 12 mm-wide tapes – determining an accurate NZPV using the above described method became tedious. It happened on occasion that the 93 K wave front did not propagate uniformly; one side of the tape quenched more rapidly than the other side, for example. In such situations it became difficult to determine if the tape really quenched or not or to what extent. To avoid such problems, in the measurements the criterion was further specified arbitrarily as "the furthest point of the 93 K zone". Apart from this, the boiling also severely distorted the measurements and the temperature rises were on occasion so quick that defining a 93 K wave front did not make a practical sense. This happened frequently in stabilizer-free tapes during a quench, where the tape was locally often either at 77 K or at some higher temperature, above the measurement limit of 260 K of the thermal imaging method.

The method is further limited in measuring the NZPV in the case of copper-stabilized HTS tapes under certain current conditions. Exceeding the critical current of the tape caused on more than one occasion the current to transfer into the copper shunt practically instantaneously. While this is the expected behaviour and exact function of the shunt layer in the first place, in such situations the HTS tapes heated almost uniformly over the length. On other occasions (mostly stabilizer-free) tapes rapidly developed film boiling over the entire surface, which then distorted the recording and as a consequence, no "thermal wave front" was visible that would allow calculating the NZPV. This was also due to current bypass into the thin silver layer, however due to the significantly lower heat capacity, the layer quickly heated up and entered a film boiling regime (refer to Figure A.3 in the Appendix). An alternative here can be the use of the boiling development over the tape's surface, however, it is unknown how this then compared to the measured NZPV as described previously.

For these reasons, while with electrical NZPV measurements it is possible to measure the quench propagation velocity at multiple current levels (even far above the tape's I_c), this is most often not the case with the described thermal imaging method.

4.2.2 Results and comparison

The results of the NZPV measurements are summarized in Table 4.1 together with the relevant properties of the tapes, where the normal zone propagation velocity could be reliably calculated from the thermal imaging as introduced. The discussion about the results is subdivided into three main groups based on the architecture of the tapes: stabilizer-free HTS tapes, HTS tapes with metallic stabilization and HTS tapes with some form of special architecture.

HTS tapes without stabilization

Results of sample 1 from Theva showed a good correlation between the thermal imaging and the electrical measurement. With the tape's critical current in the range of 600 A to 700 A, 650 A and 625 A pulses for a duration of 12 ms have both recorded a normal zone propagation velocity of $5\,\mathrm{m\,s^{-1}}$.

Sample 2 was practically identical to sample 1, however, this was damaged during the pulsed measurements at KIT. Here a 5-fold lower NZPV was measured at merely 63 % of the critical current.

Sample 6 was a tape from SuperPower with only an Ag-layer. With a current pulse amplitude of 158 A, an NZPV of $1.82\,\mathrm{m\,s^{-1}}$ was measured electrically. With the thermal imaging method a normal zone propagation was only observed much higher at 180 A (over 60 % above the critical current) with a mere $0.83\,\mathrm{m\,s^{-1}}$. This is about half that of the electrical measurement result and the exact reason is unknown.

Table 4.1: NZPV measurement results using fluorescent thermal imaging and comparison with electrical measurements as reported by Polytechnique Montréal for the same or similar samples. † Measured at 77 K, self-field. ‡ Pulse amplitude in the electrical and optical measurements for which the NZPV is reported. Samples 3, 4, 5, 8 and 9 are discussed in greater detail in Section 4.3.

No.	Manufacturer	Width (mm)	Finish	† I_c (A)	‡$\text{Pulse}_{\text{el.}}$ (A)	$\text{NZPV}_{\text{el.}}$ (m s^{-1})	‡$\text{Pulse}_{\text{opt.}}$ (A)	$\text{NZPV}_{\text{opt.}}$ (m s^{-1})
1	Theva	12	Ag	600 to 700	650	5	625	5
2	Theva	12	Ag	637	650	5	400	0.29 to 0.7
3	Theva	12	Ag with 100 µmCu-Ni	600 to 700	600	0.54	600	0.2 to 0.5
4	SuperPower	4	Ag	80 to 90	85	0.3	115	0.59 to 0.71
5	SuperPower	4	Ag (CFD)	80 to 90	85	2.2	85	-
6	SuperPower	4	Ag	110 to 120	158	1.82	180	0.4 to 0.83
7	SuperPower	12	Ag	331	330	1.96	330	0.17
8	Oxolutia	12	Ag (CFD)	700 (at 47.9 mT)	-	-	650	7.5
9	SuperOx	12	Ag with 20 µm Cu	331	-	-	400	0.0522

A similar mismatch was observed in the electrical and thermal measurement results of sample 7 from SuperPower. Pulsing the tape with 100 % of its critical current resulted in an NZPV of 1.96 m s^{-1} in the electrical measurement and only 0.17 m s^{-1} in the thermal imaging. This constitutes an order of magnitude difference in the two measurement methods.

Sample 4 from SuperPower was interesting, since while with the electrical measurement method an NZPV was determined for a current pulse of 85 A as 0.3 m s^{-1}, at this current level the tape did not show any signs of a quench when measured at KIT. A quench was observed only at 115 A first with a normal zone propagation velocity of 0.71 m s^{-1}.

HTS tapes with stabilization

Sample 3 was a Theva sample with a nickel-copper alloy as a stabilization. The NZPV was measured as 0.5 m s^{-1} at both institutes when pulsed with 600 A (I_c), showing a close match between the two methods.

Sample 9 was a sample from SuperOx with a standard 20 µm copper stabilization. Applying a current pulse 20 % above the tape's I_c resulted in an NZPV of 0.5 m s^{-1}, however, data from Montréal was unfortunately not available for this tape.

HTS tapes with special architecture

Two tapes with special, current flow diverter architectures [110] were also measured (this type of tape architecture is introduced in greater detail in Section 4.3.3). Sample 5 was prepared from a sample practically identical to sample 4 by the group at Montréal. While the electrical NZPV measurement showed a quench propagation speed of 3.3 m s^{-1}, an over 7-fold improvement, with respect to sample 4, thermal imaging could not reliably determine an NZPV. In the measurements, the tape quickly developed a boiling region over the entire surface and as a consequence no travelling wave front was observed. It is believed, however, that this is the sign of a considerably improved NZPV as previously discussed in Section 4.2.1.

Another current flow diverter sample was prepared by Oxolutia with a novel process using inkjet printing [111, 112]. A sample was measured successfully – for the first time – at KIT for NZPV with the fluorescent thermal imaging method. A speed of around 7.5 m s^{-1} was obtained at approximately 93 % of the critical current. This sample is discussed in more detail in Section 4.3.

4.3 Qualitative quench investigation

The promise of high-speed fluorescent thermal imaging is the additional, qualitative information provided about the quench development aside the capability of calculating the NZPV. This aspect of quench is not often researched and therefore constitutes the focus point of this section.

Here the samples 3, 4, 5, 8 and 9 from Table 4.1 are discussed in greater detail with respect to the quench formation and development. Electrical measurements are shown for different current pulses together with both thermal and raw fluorescent images. These latter are often valuable in recordings where liquid nitrogen boiling considerably distorted the image. In such cases, the quench formation is better visible on the raw images than on the post-processed thermal images.

4.3.1 HTS tape without stabilisation

Here the electrical and thermal imaging results of a 4 mm-wide, 120 mm-long HTS sample from SuperPower are presented on the example of sample 4 in Figure 4.7. The stabilizer-free sample had solely a 1.2 µm thick Ag coating, making it ideal to show a characteristic quench propagation of such a tape. The I_c of the sample was 85 A, however, at current pulses up to 110 A no quench was visible on the thermal imaging.

The first six current pulses in Figure 4.7 show clearly visible voltage signals, but again, no quench on the thermal imaging. The pulse duration was kept constant at 10 ms and the current amplitude was increased from 85 A to 110 A. The absorbed energies

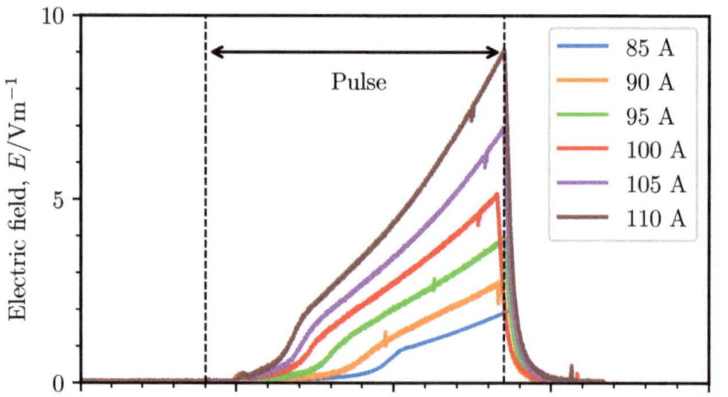

Figure 4.7: Sample 4 pulsed with 85 A to 110 A, 10 ms long pulses. Clear voltage signals are visible, however, the thermal imaging showed no signs of a quench.

were 1.53 J, 1.62 J, 1.71 J, 1.8 J, 1.87 J and 1.98 J, respectively, calculated from eq. 4.1. Electric fields of up to 9 V m^{-1} were recorded that are in the range where the electrical signals are used for NZPV calculation.

The current amplitude was increased further and a quench became first visible at 115 A (135 % I_c). As indicated in Figure 4.8a, on the thermal imaging the quench is first visible at 12.65 ms where the measured electric field was 6.3 V m^{-1}. The extracted images in Figure 4.8b-4.8d show uniform quench propagation along the width of the tape and eventually heating to beyond 260 K. Note the gradual voltage development across the sample over the time of the pulse. Calculating the total deposited energy based on eq. 4.1 results in 4.36 J, whereas the absorbed energy until a visible quench was merely 0.57 J.

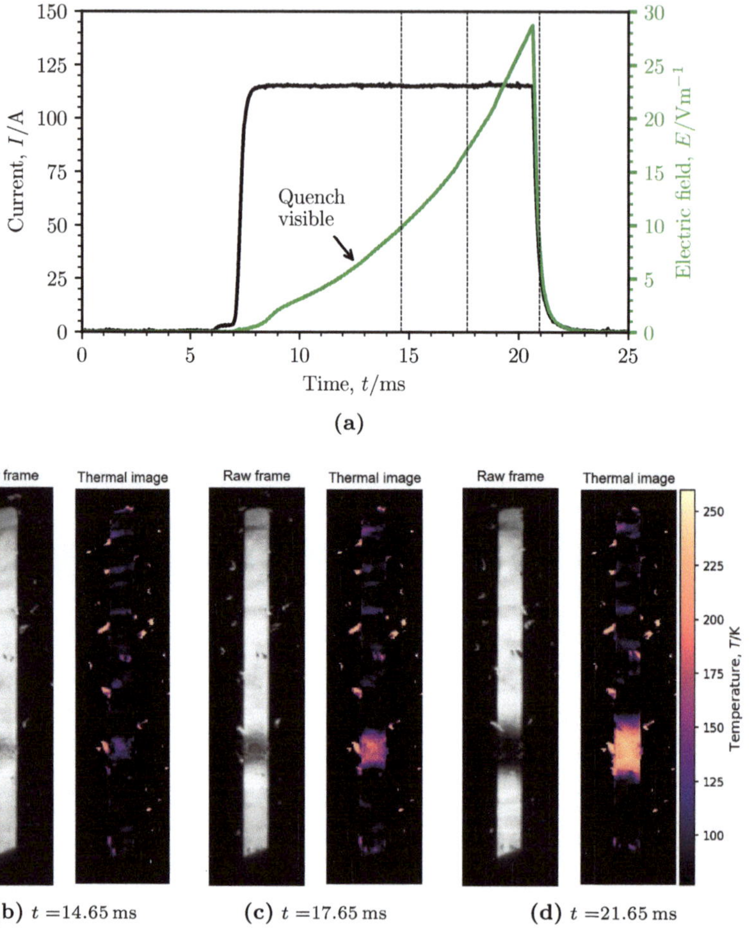

Figure 4.8: Sample 4 pulsed with a 115 A, 12 ms long pulse. (a) shows the applied current pulse and the recorded electric field across the tape. (b)-(d) show raw and thermal images corresponding to times indicated by the black vertical lines in (a).

4.3.2 HTS tapes with stabilisation

Theva tape with 100 µm Cu-Ni

Sample number 3 was prepared by Theva, using a special, 100 µm copper-nickel electrical shunt (refer to Table 4.1) and the measurements on this sample show well the general behaviour of metal-stabilized tapes. In the following, three separate measurements are discussed where different current pulses were applied and, accordingly, the tape quenched in different ways. These were the following

- 700 A (~100 % I_c) for 12 ms
- 900 A (130 % I_c) for 12 ms
- 600 A (~100 % I_c) for 100 ms.

The shortcomings of the thermal imaging method are clearly visible – as it often happened with metal-stabilised samples – where the current bypass rapidly heated up the tape surface and formed a film boiling layer. This layer then partially or completely prevented obtaining temperatures over the surface, however, the raw images still showed the general quench behaviour.

Figure 4.9 shows the sample's behaviour when pulsed with the 700 A, 12 ms long pulse. Unfortunately, the quench is only visible on the raw images. Here a uniformly propagating quench region is visible, starting at the magnetic defect's location. It is also clear that the quench has only propagated about ~0.5 cm in total as shown in Figure 4.9d. The energy deposited into the tape was also calculated according to eq. 4.1, which yielded 2.09 J.

The sample was subsequently pulsed with 900 A for the same 12 ms duration as indicated in Figure 4.10a. At this current level, the voltage drop across the sample became too high and the current source went into voltage limiting, causing a drop in the output current. Nevertheless, after the initial quench zone formation at the magnet's position, several additional quench zones became visible, as shown in Figure 4.10d. The appearance of multiple quench zones – here and in other measurement as well – is hypothesised by the presence of material defects, causing local drops in the tape's J_c and thereby facilitating further quench nucleation after the initial transition began. The energy deposited in the tape during this pulse was 19.64 J, over the same 12 ms pulse duration.

In a third measurement a lower, 600 A pulse was applied but for longer, 100 ms. As shown in Figure 4.11a, the current source was yet again driven into voltage limit and the current across the sample dropped. However, this time the sample only quenched at the magnetic defect and the quench propagated from there. No multiple quench zones

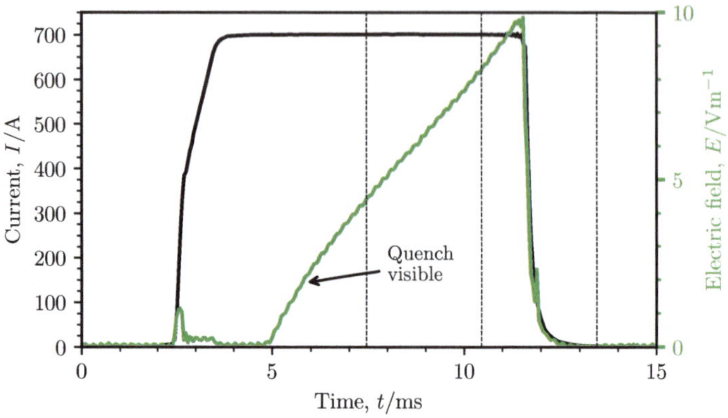

(a)

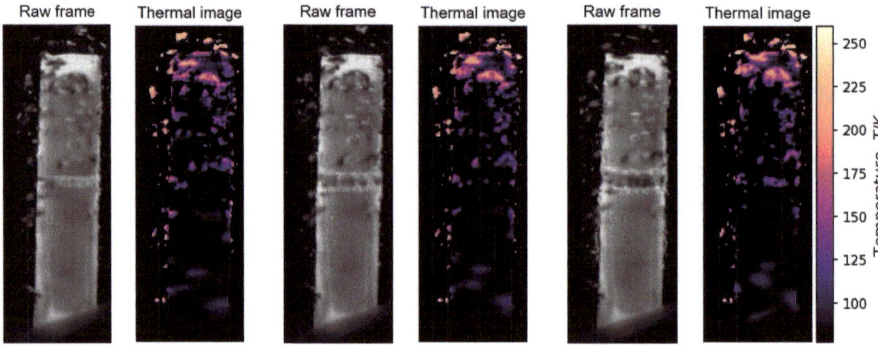

(b) t =7.45 ms (c) t =35.45 ms (d) t =13.45 ms

Figure 4.9: Sample 3 pulsed with a 700 A, 12 ms long pulse. (a) shows the applied current pulse and the recorded electric field across the tape. (b)-(d) show raw and thermal images corresponding to times indicated by the black vertical lines in (a).

were observed in this experiment and the thermal imaging showed heating to over 260 K, due to an overall deposited energy of 51.42 J. This energy is over twice as much as in the previous experiment, however it was absorbed over the duration of 100 ms.

The three measurements would seem to indicate that a current pulse close to the I_c of the conductor would create a quench at the weakest point in the tape, at the magnetic defect in this case. If the pulse duration is short, the normal zone simply does not have enough energy (or time) to propagate. Increasing the pulse duration provides this additional energy and hence causes a propagating quench zone. On the other hand, if a current pulse considerably above the tape's critical current were applied, quench zones appeared at all (hypothesised) defects in the conductor and spread the energy out over the entire surface.

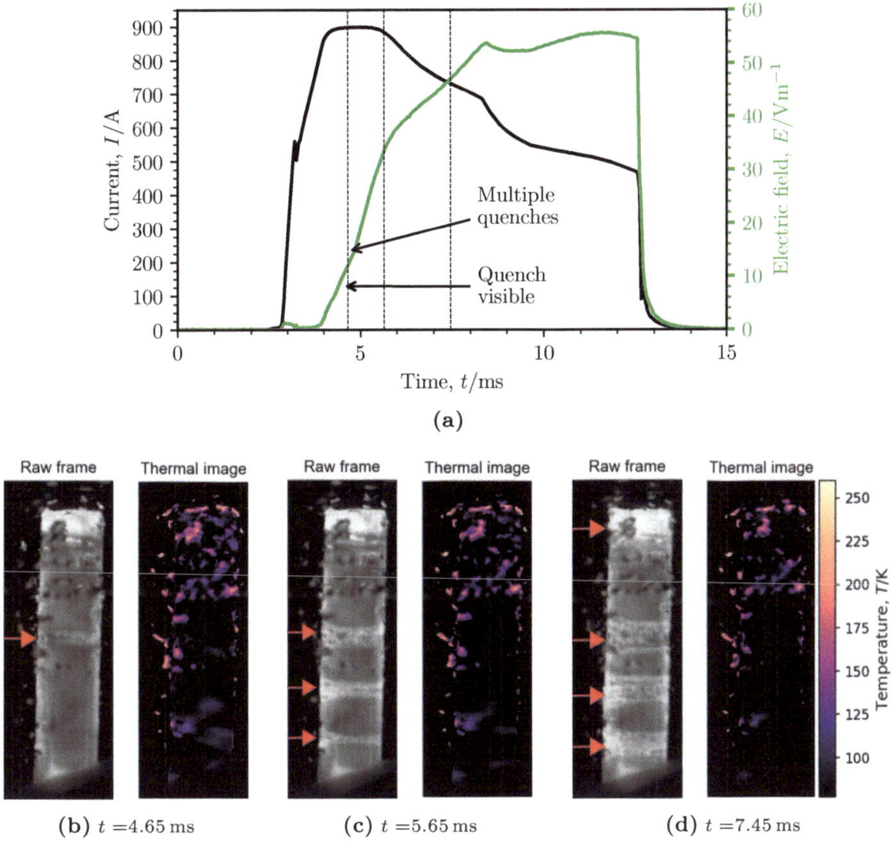

Figure 4.10: Sample 3 pulsed with a 900 A, 12 ms long pulse. (a) shows the applied current pulse and the recorded electric field across the tape. (b)-(d) show raw and thermal images corresponding to times indicated by the black vertical lines in (a). The red arrows indicate the quench nucleation positions.

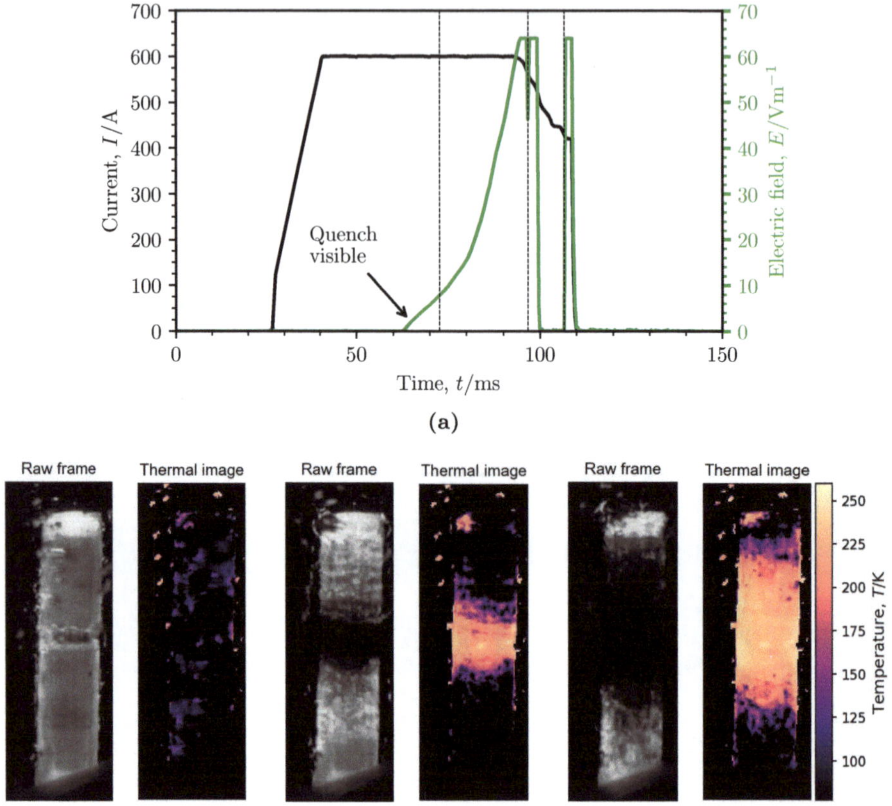

(a)

(b) t =72.67 ms (c) t =94.67 ms (d) t =106.67 ms

Figure 4.11: Sample 3 pulsed with a 600 A, 100 ms long pulse. (a) shows the applied current pulse and the recorded electric field across the tape. (b)-(d) show raw and thermal images corresponding to times indicated by the black vertical lines in (a).

SuperOx tape with 20 µm Cu

Similar behaviour was observed on sample 9 from SuperOx with a 20 µm pure copper stabilization. The tape was pulsed with

- 500 A (150 % I_c) for 50 ms
- 700 A (210 % I_c) for 12 ms.

Figure 4.12 shows the 500 A pulse applied to the tape for 50 ms. While initially a small quench zone appeared, Figure 4.12b, quickly multiple quench zones developed as shown in Figures 4.12c and 4.12d. This is again hypothesised to happen since the tape was pulsed with a high, $1.5 I_c$, pulse. Due to the longer, 50 ms pulse duration all of the initiated quench zones began spreading and formed a film boiling over the surface. This again indicates that current pulses above the conductor's critical current cause quench nucleation at several locations, possibly material defects. Calculating the total energy deposited in the tape during the pulse shows that 8.45 J were absorbed.

In the second measurement the tape was pulsed with an even higher amplitude, but for a shorter duration, 700 A for 12 ms. Here the deposited energy during the pulse was 7.02 J, about 15 % less than in the previous measurement. However, as shown in Figure 4.13, the tape behaved completely differently. Again, the shortcoming of the thermal imaging method is obvious: it is impossible to obtain temperatures during the quench as the heavy surface boiling completely distorts the image. Regardless, the raw fluorescent images can be relied on. Only a single quench zone appeared, and quickly thereafter the entire surface of the tape developed a film boiling. It is furthermore interesting – although difficult to see on static images – that even though the quench initially started at the magnetic defect, it quickly distributed along the edges over the length of the tape and thereafter formed the film boiling. Applying currents far above the critical current caused the current to transfer rapidly in the metallic shunt, where due to its finite resistance this heated up, and caused surface boiling over the entire surface of the tape, distributing the energy.

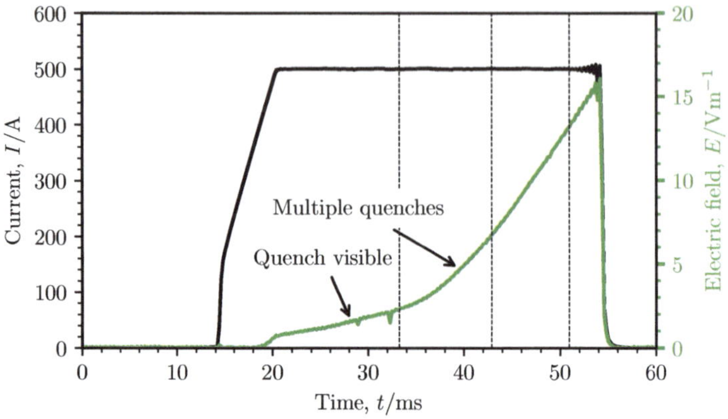

(a)

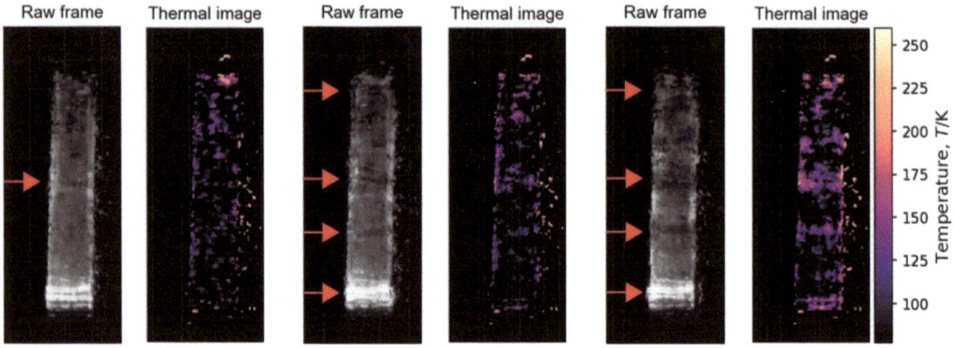

(b) t =33.25 ms **(c)** t =42.92 ms **(d)** t =50.92 ms

Figure 4.12: Sample 9 pulsed with a 500 A, 50 ms long pulse. (a) shows the applied current pulse and the recorded electric field across the tape. The arrows show the point where the quench is first visible on the thermal imaging recording and where multiple quench zones started forming. (b)-(d) show raw and thermal images corresponding to times indicated by the black vertical lines in (a). The red arrows indicate the quench nucleation positions.

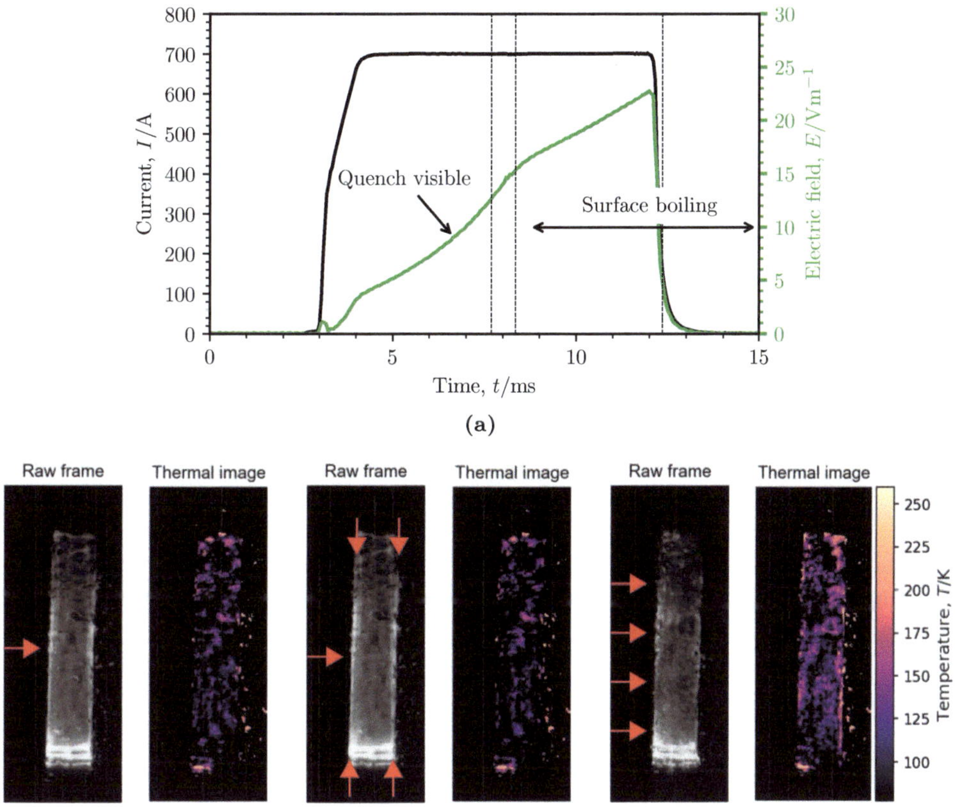

(b) $t = 7.68\,\mathrm{ms}$ **(c)** $t = 8.35\,\mathrm{ms}$ **(d)** $t = 12.35\,\mathrm{ms}$

Figure 4.13: Sample 9 pulsed with a 700 A, 12 ms long pulse. (a) shows the applied current pulse and the recorded electric field across the tape. The arrows show the point where the quench is first visible on the thermal imaging recording and where multiple quench zones started forming. (b)-(d) show raw and thermal images corresponding to times indicated by the black vertical lines in (a). The red arrows indicate the quench development.

4.3.3 Tapes with current flow diverter architecture

A special application of fluorescent thermal imaging can be the qualitative inspection of quench in HTS tapes with special or experimental architectures, aimed at e.g. accelerating the NZPV. A so-called "current flow diverter" (CFD) architecture was proposed by Lacroix et al. [110] with this exact purpose. The principle is that by creating an area of high inter-facial resistance between the superconducting layer and the upper silver stabilisation along the centre of the tape, see Figure 4.14b, in case of a quench the current is forced to flow along the edges, a lower resistivity path. The increased current density in this smaller area causes local heating and propels the quench quickly along the edges.

An alternative implementation of the CFD architecture is called b-CFD [47], where no high-resistivity buffer layer is necessary between the superconducting layer and the Ag coating (Figure 4.14c). Instead, the thickness of the silver stabilization on the surface of the tape (superconductor side) is reduced, and simultaneously increased at the back (Hastelloy® side). Then by adding a thin conducting bypass around the longitudinal edges of the tape (e.g. silver) the current during quench can be forced to transfer via the edges to the bottom of the tape, a path with lower resistance. This, in effect, achieves the same goal of accelerating the quench propagation via the low heat capacity electrical conductor along the edges of the tape.

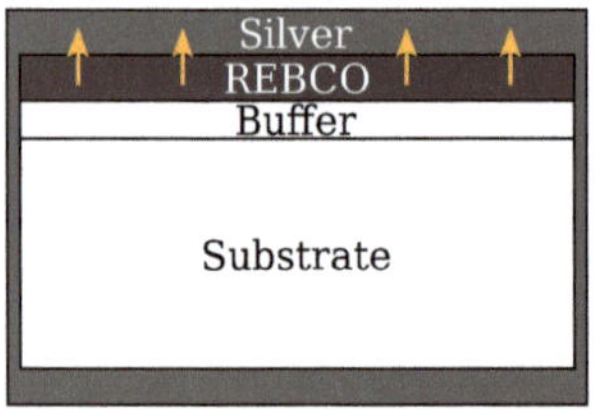

(a) Stabilizer-free HTS tape.

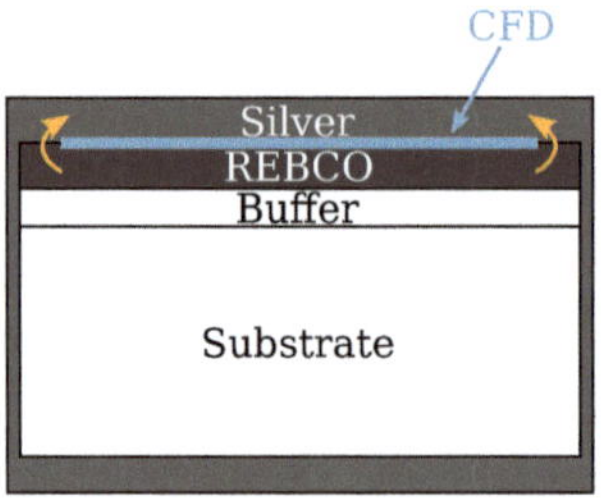

(b) HTS tape with CFD layer.

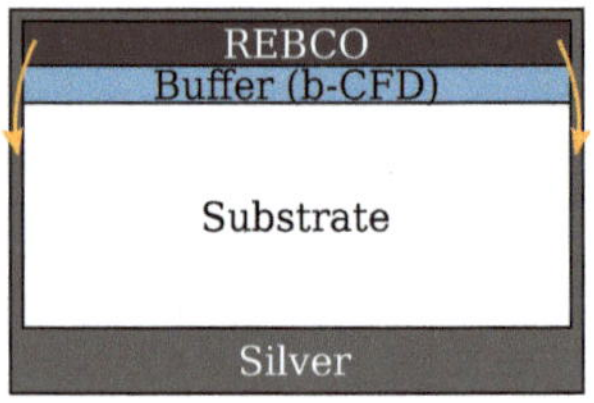

(c) HTS tape with b-CFD architecture.

Figure 4.14: Cross-section of various HTS tape architectures. Subfigure (a) shows a common HTS tape, with uniform inter-facial resistance, (b) shows the CFD architecture with an artificially increased inter-facial resistance and (c) shows the b-CFD concept where the thickness of the Ag stabilization is changed on the front- and back-side of the tape [47]. The arrows indicate current sharing paths.

Inkjet printed CFD

Oxolutia is a company based in Barcelona, Spain, developing a novel method to introduce a CFD architecture in HTS tapes. Their patented approach [113] is to use inkjet printing to apply a highly resistive interface layer between the superconductor layer and the silver finish [111, 112]. A sample was prepared by Oxolutia, using a Theva tape as a basis, with a 100 µm thick substrate. On top, 3.5 µm MgO and a 3.1 nm thick GDBCO layer were deposited. This was then followed by inkjet printing a ~110 nm thick Y_2O_3 layer over 90 % of the width (centred), leaving the sides uncovered. The tape was then stabilized by sputtering 1 µm Ag on the surface. As a result, the tape edges have a low contact resistance between the superconducting and the Ag layer. Simultaneously, the Y_2O_3-covered part of the tape has a high interface resistance and achieves the CFD architecture similarly as shown in Figure 4.14b. Several shorter, 6 cm, samples were prepared with the above manufacturing method and were tested at KIT to determine both the NZPV as well as to observe whether the tape behaves differently during quench. Unfortunately, the high interface resistance – together with the short length – caused problems when injecting current in the samples. Often the contact resistance was either too high to force current in the sample with the available current supply or on other occasions the superconductor layer burned out under the current contacts. No reliable measurements could be done on such short samples. The first successful measurement was carried out on a longer, 10 cm sample. The I_c distribution of this sample was measured at the Institut de Ciència de Materials de Barcelona (ICMAB) using Scanning Hall Probe Microscopy [114]. The sample was found to have an I_c of over 700 A in a magnetic field of 47.9 mT.

Since this was a prototype HTS sample without any metallic stabilization, 12 ms current pulses were applied where the current amplitude was gradually increased. The sample eventually did quench at a current amplitude of 650 A, see Figure 4.15a, and a peak NZPV of 7.5 m s^{-1} was measured. The reason it is referred to as "peak NZPV" is since the current got limited in the sample at around 11 ms, after which the quench propagation slowed down. The electric field appeared suddenly across the sample and saturated the amplifier's 1 V range. However, because the current across the sample dropped, the current source must have gone into voltage limiting due to the high resistance across the sample, which happens at ~10 V. Over the distance of the voltage taps in this measurement, the electric field across the sample was, at least, 179 V m^{-1}. The appearance of the voltage signal matched exactly the visual appearance of a quench in the sample.

The sample also revealed an interestingly-shaped quench zone. Looking at Figure 4.15c a hot zone is visible around the centre of the tape, where the magnet is located. In addition, two further quench zones are visible at either side of the magnetic defect, at the edges of the tape. This could very well be caused by the CFD architecture, where after the initial quench at the magnetic defect the current was diverted to a lower resistivity region around the edges of the tape and caused rapid heating. The sample was measured after the initial quench and no damage to the superconducting layer was discovered. This would seem to indicate that the prepared sample is on track to meet the requirements set forth by the FASTGRID project, aiming at developing HTS conductors that can withstand electric fields beyond $100\,\mathrm{V\,m^{-1}}$ without damage [107]. Further measurements on closely identical samples are currently ongoing at collaborating institutes to measure the NZPV electrically as well.

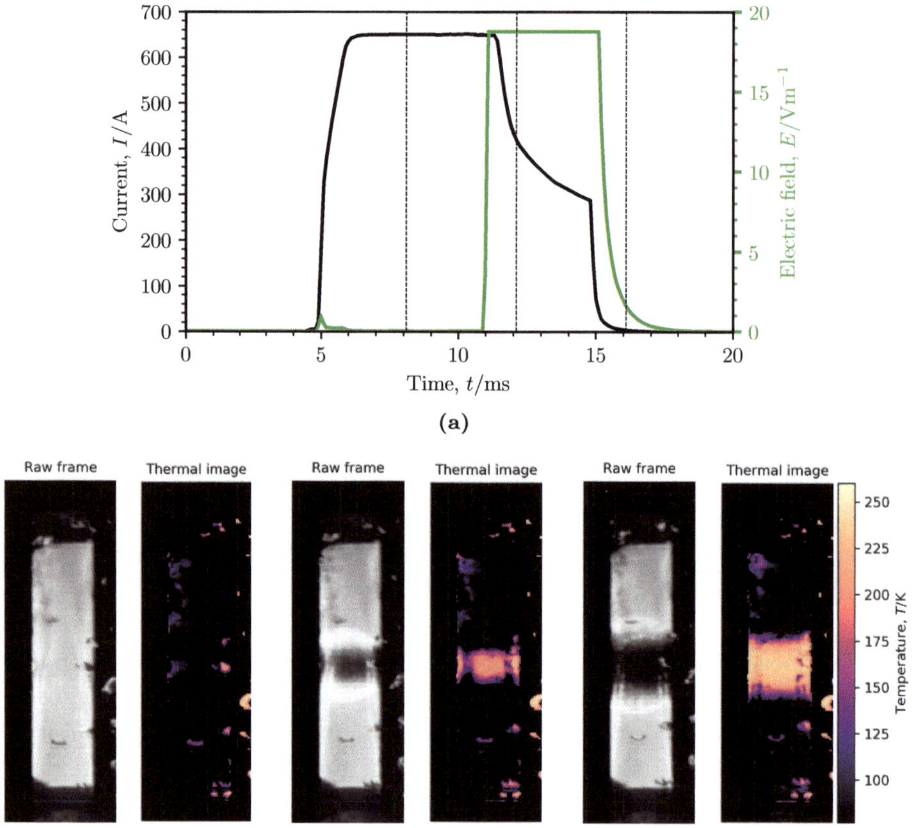

(a)

(b) t =8.1 ms (c) t =12.1 ms (d) t =16.1 ms

Figure 4.15: Sample 8 pulsed with a 650 A, 12 ms long pulse. (a) shows the applied current pulse and the recorded electric field across the tape. (b)-(d) show raw and thermal images corresponding to times indicated by the black vertical lines in (a).

CFD architecture at Polytechnique Montréal

The measurement results of sample 5 are shown in Figure 4.16. It was prepared at Polytechnique Montréal with a CFD architecture [18] using a conductor similar to sample 4 as a basis and sent to KIT for NZPV measurement and for comparing results. Having an I_c of 85 A, this tape was pulsed with 85 A for a duration of 8 ms, where an NZPV of ~2.2 m s^{-1} was measured at Polytechnique Montréal (see Table 4.1). Reproducing the same pulse at KIT using high-speed fluorescent thermal imaging did not yield the same results, however. While a clear electrical quench signal was visible as shown in Figure 4.16a, the thermal images did not capture a propagating quench zone. The only visible sign of the quench was the rapid development of a thin film boiling over the entire surface of the tape.

This behaviour is similar to that observed in metal stabilized tapes as previously discussed, where in the case of over-critical current pulses the current would divert into the electrical shunt layer and cause a rapid film boiling layer development. Here, however, there was no metallic shunt layer, only the standard 1.2 µm thick Ag coating together with the CFD architecture. Regardless of this, no reasonable NZPV value could be obtained for this tape with the thermal imaging method.

There is no clear explanation for the behaviour of this Ag tape. It is believed based on the existing experimental data, however, that the CFD architecture indeed forces the current to divert rapidly to the edges of the conductor, where it encounters a low resistivity path. Here it can flow again across both the length of the conductor and across the width of the superconductor face. This can result in a rapid surface boiling and in an almost instant, simultaneous surface boiling as opposed to a propagating zone, where the current can locally transfer from the superconducting layer into the silver layer directly above it. The results indicate that the CFD architecture does help distribute the heating over the tape's length and as a result it can be considered as a viable method for quench protection in HTS coated conductors.

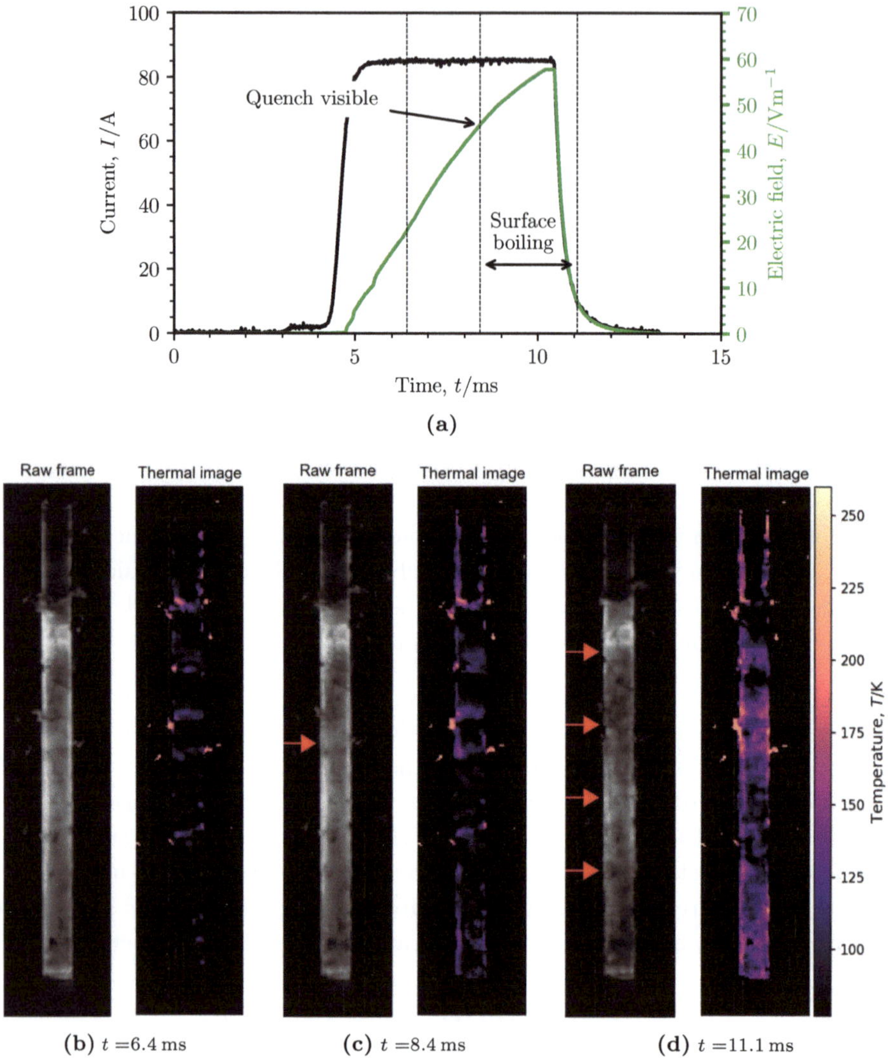

Figure 4.16: Sample 5 (with a CFD architecture added at Polytechnique Montréal [18]) pulsed with a 85 A, 8 ms long pulse. (a) shows the applied current pulse and the recorded electric field across the tape. (b)-(d) show raw and thermal images corresponding to times indicated by the black vertical lines in (a). The red arrows indicate the quench development.

4.4 Summary

4.4.1 NZPV calculation validation

This chapter demonstrated that the normal zone propagation velocity can be obtained via the introduced thermal imaging method purely from the developed temperatures over the surface of an HTS tape. While in electrical NZPV measurements the quench propagation is calculated as the velocity of the developing electric field, in the shown thermal imaging approach the quench was defined as "the furthest point of the 93 K wave font". Consequently, the velocity of this wave front was calculated to determine the NZPV from a thermal perspective. Due to the development of film boiling regions and the inherent differences compared to the electrical NZPV measurement method, the shown thermal imaging cannot be used to calculate an NZPV in every case. The method is most suited in this respect to investigations until approximately 150 $\% I_c$.

Measurement results on several HTS samples were presented with varying architecture in terms of substrate thickness, silver layer thickness as well as copper stabilization. Unfortunately data was not always available from both electrical and thermal imaging NZPV measurements and as a consequence the comparison is sparse. Nevertheless, several measurements indicate a reasonable match between the two NZPV measurement approaches, using different physical principles. For two Theva tapes, one stabilizer-free and another with a copper-nickel shunt both methods reported comparable NZPV values at closely identical current pulses. In another Theva tape measurement the sample quenched at a lower current level in the optical measurement and hence a lower NZPV was obtained. In general in the case of not matching results, the thermal imaging tends to under-estimate the NZPV reported by the electrical measurement method. Two such prominent exceptions were recorded as well, in stabilizer-free samples from SuperPower. These have shown a rather large discrepancy in the results between the two methods with a factor of 2 to 10 times slower NZPV measured by the thermal imaging. This was expected to a certain extent as the heating is a consequence of a developed electric field in the tape, and as such the temperature rise must lag a voltage signal. Simultaneously, the discrepancies between the two methods were observed in tapes with a critical current roughly half when compared to measurements where the two methods reported similar results. Since losses (and therefore heating) are proportional to the second power of the applied current, this may suggest that the thermally and electrically measured NZPVs are similar at higher current amplitudes due to the higher heating power during a quench. This also applies to a lower current amplitude, where a lower electric field may develop faster, followed by a slower heating due to a lower power dissipation.

Measuring the NZPV using fluorescent thermal imaging has its own limitations. Obtaining an NZPV for different current levels was difficult, especially for currents significantly above the I_c of the respective conductor. Using high current amplitudes the current rapidly transfers to the metallic shunt (if present) and causes quasi-uniform heating over the entire observed surface, thereby preventing the formation of a propagating normal zone to be observed. In stabilizer-free tapes, on the other hand, a thin film boiling appeared over the entire surface at such high current levels, which then distorted the view and prevented the calculation of an NZPV. While the method is capable of reporting the normal zone propagation velocity in HTS tapes, it was not aimed primarily for this investigation. The method furthermore requires considerable manual post-processing that cannot be completely automated. As a result, while the thermal imaging method can serve as a comparison for NZPV measurement, it is not intended to replace other methods.

4.4.2 Effect of metallic stabilization

In Section 4.3.2 samples 3 and 9 with the 100 µm Cu-Ni and the 20 µm pure copper shunt, respectively, were discussed. It was shown that although the thickness of the metallic stabilization, as well as the composition, were different, the tapes quenched in a similar way with respect to the applied current pulse amplitude and duration. This is not to say that the NZPV or the stability of the tapes is independent of the metallic shunt thickness, rather that the qualitative appearance of quenched zones was observed in both tapes in a similar fashion.

4.4.3 Electric field development

When comparing the electrical measurement results of the Oxolutia sample with the CFD architecture from Figure 4.15a (sample 8) with that of e.g the SuperPower sample shown in Figure 4.8 (sample 4), the difference in the voltage profiles is immediately obvious. While the SuperPower sample developed the voltage gradually over the duration of the pulse, the Oxolutia sample answered with a step-like response to the applied current pulse. At the time of writing this was the first successful measurement on such a tape architecture and as such there was no data available from other institutes to confirm the reproducibility of the quench behaviour. The most likely explanation is related to the presence of the high inter-facial resistance between the superconducting and the covering silver layer. With a good electrical connection between these layers (in conventional HTS tapes), as soon as the superconductor's J_c is approached locally, a voltage can be recorded across the tape. Using a highly resistive CFD layer presents a barrier to current transfer into the silver layer, essentially forcing the current to stay in the superconductor. This could then cause $J_c(T)$ to drop rapidly inside the superconductor, due to a tiny, dissipative zone and consequently cause a high electric field in accordance with the power law (Chapter 2). At a high enough voltage then current sharing between the silver and superconducting layer will take place and cause rapid local heating. In the shown measurement the electric field became so high that the current source could not maintain the required voltage to supply the requested 115 A.

Furthermore, on the example of the same SuperPower sample, shown in Figure 4.8, one can see that a relatively high electric field of up to $9\,\mathrm{V\,m^{-1}}$ was recorded, however no quench was visible on the thermal imaging. Simultaneously, at this "not quenched" current pulse of 110 A an energy of 1.98 J was absorbed by the tape. Then in the measurement where the tape did quench at 115 A, merely 0.57 J was absorbed over the first ~7 ms of the pulse (until the appearance of the quench) and the electric field was $6.3\,\mathrm{V\,m^{-1}}$. This proves the existing difficulty that measuring an electric field over a longer distance does not necessarily give accurate information about the development of a localized dissipative zone. In general, a quench could be identified on the thermal images at an approximate electric field of $5\,\mathrm{V\,m^{-1}}$, however, this was highly dependent on the distance between the voltage taps.

On occasion, HTS tapes did burn out during measurement if the deposited energy was too high or concentrated in a single point. In most situations a burnout could be easily identified on the electrical signal, without inspecting the tape. In such cases the resistance of the tape section measured in between the voltage taps became too large, which resulted in a large, sharp voltage drop. As the maximum voltage of the current supply is limited, the current would consequently fall to a sustainable level. Such a measurement signal, especially with stabilizer-free tapes, indicated that the sample (most likely) suffered permanent damage. Tapes with sufficient metallic stabilization on the other hand may survive such a scenario due to their increased heat capacity. It was therefore surprising to observe that in the case of the Oxolutia sample (number 8) with the inkjet-printed CFD – despite the lack of a metallic stabilization – the sample did not suffer any damage even though the developed electric field was high enough to cause a drop in the supplied current.

4.4.4 Types of quench development

As shown in previous measurements, a quench can behave in several different ways. Probably the biggest distinction that can be made is how the quench spreads: starting from a single point or appearing at several locations almost simultaneously. Unfortunately, it was not possible to measure the NZPV in the current experimental setup simultaneously using both the optical and electrical methods. However, since the

NZPV of HTS is routinely measured at other institutes for a range of current levels (including over-I_c) most likely the voltage profile during such a quench with multiple nucleation points is indistinguishable from that of a “normal”, propagating quench zone. From the thermal images and results presented here, at the same time, it is clear that in such scenarios the heat distribution happened vastly differently.

As a conclusion, one can say that it is not only the deposited energy in the tape that is important, but how or rather what type of current pulse was applied, and consequently the rate at which the energy was deposited (heating power). It is already known that the NZPV of HTS is higher at higher current levels with respect to I_c [16], however the qualitative formation of normal regions was not investigated in detail so far.

N. Nguyen et al. [115] previously discovered non-homogeneous “banding” boiling over HTS and also showed that currents far above the tape's I_c caused a more homogeneous quench. Conclusions similar to this were drawn in this work as well and these can be summarized as follows. At “low” powers, corresponding to low amplitude current pulses, HTS tapes developed a single quench zone and depending on the duration of the pulse this zone then propagated. At “intermediate” heating rates, ~30 % to 50 % above the I_c, coated conductors seemed to develop several quench zones almost simultaneously that rapidly expanded and caused heating and boiling of the coolant over the tape's (almost entire) surface. Then at even higher heating power and pulse amplitudes, the current transferred almost instantly into the metallic shunt and caused Joule heating over the entire tape surface.

This is an interesting phenomenon worth further experimental investigation. By designing HTS tapes with shunt layers having optimal heat capacity and resistivity characteristics, they can be fine-tuned for specific applications (e.g. superconducting fault current limiters). By adjusting these parameters, several simultaneous quench zones could be caused at the same time and if the temperature rise is kept in control, the tape could be held in the nucleate boiling regime with optimal heat transfer conditions (refer to Appendix, Figure A.3). In the current understanding, a quench nucleates in an HTS at a single point and spreads with a given NZPV along the length. This is supported by the standard electric NZPV measurements. In light of the appearance of multiple quench zones presented in this work as well, the quench and NZPV may not be defined as straightforward as previously thought. Although the NZPV of HTS tapes is undoubtedly not fast enough, by the appearance of multiple quench nucleation points, the energy could still be distributed and a thermal runaway prevented.

5 Fluorescent thermal imaging of HTS pancake coils

Previous chapters described the working principles of the fluorescent thermal imaging method when combined with high-speed recordings and also showed its application in quench investigation and NZPV measurement of single HTS tapes. This chapter shows the application of the method on the example of two distinct pancake coils wound from 2G HTS tapes: a conventional insulated and a non-insulated coil. The chapter details the design and preparation of the coils as well as the experimental setup, sets the aims for their measurement and gives a comparison of their final performance and behaviour in operation.

5.1 Experimental objectives

The stability and quench aspects of HTS coils were introduced previously in Chapter 2. As a proof of concept, fluorescent thermal imaging was implemented for measuring two HTS pancake coils, wound with different winding techniques. The first coil was a conventional insulated coil, whereas the second coil was wound without any turn-to-turn insulation.

Both coils have their own benefits and drawbacks and the aim of the investigation was to observe the coils during operation from a thermal and electrical perspective. Such a thermal imaging on insulated or NI coils was not carried out in the past and was not commonly available using state-of-the-art measurement methods. With the thermal imaging possible weak spots can be detected in the coil, coming either from design, material defects or simply due to exceeding the critical current of the coil locally due to some disturbance.

5.2 Electro-magnetic model

A number of numerical models exist using finite element modelling (FEM) and finite difference modelling (FDM) as well as equivalent circuit models [73, 86, 116–120] for simulating the behaviour and calculating the critical current of superconducting pancake coils including NI coils.

The development and simulation of a complete coil was beyond the scope of this work, however, FEM was used as initial analysis to determine the expected critical current, magnetic field as well as inductance and stored energy in the coils. A 2D axisymmetric, magneto-static model of the coil was created in the commercial, finite element method COMSOL Multiphysics 5.4 software, using the modified load line method [121]. It was implemented using COMSOL's magnetic field module, which uses the magnetic vector potential **A** as a state variable and the governing equations are

$$\mathbf{J} = \nabla \times \mathbf{H} \tag{5.1}$$

$$\mathbf{B} = \nabla \times \mathbf{A} \tag{5.2}$$

$$\mathbf{J} = \sigma \mathbf{E}. \tag{5.3}$$

In the model the HTS windings were simulated as homogenized superconducting domains of concentric rings as shown in Figure 5.1.

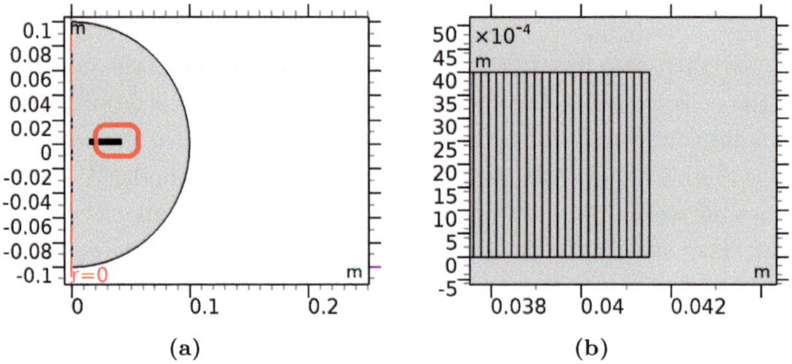

Figure 5.1: (a) shows a cross-section view of the coil turns and the air domain around it, where r is axis of symmetry. (b) shows a zoomed view of the of the coil geometry as highlighted by the red rectangle in (a), where each thin strip represents a turn of the pancake coil.

The FEM model takes the electrical conductivity σ_{sc} as input parameter, which was defined using the power law from e.q. 2.3. Furthermore, to avoid numerical instabilities – such as infinite or zero electrical conductivity – the final electrical conductivity of the superconducting domains was defined as

$$\sigma_{\mathrm{sc}} = \begin{cases} \dfrac{J_{\mathrm{c}}(B,\Theta)}{|E|}\left(\dfrac{|E|}{E_{\mathrm{c}}}\right)^{\frac{1}{n(B,\Theta)}} + \sigma_{\mathrm{norm}}, & \text{if } \sigma_{\mathrm{sc}} < \sigma_{\mathrm{max}} \\ \sigma_{\mathrm{max}}, & \text{otherwise.} \end{cases} \quad (5.4)$$

Here $E_{\mathrm{c}} = 1 \times 10^{4}\,\mathrm{V\,m^{-1}}$, $\sigma_{\mathrm{norm}} = 1 \times 10^{6}\,\mathrm{S\,m^{-1}}$ is the normal state conductivity and $\sigma_{\mathrm{max}} = 1 \times 10^{16}\,\mathrm{S\,m^{-1}}$ is an artificial maximum limit for the conductivity of the superconductor. For simplification, adiabatic conditions were assumed at a temperature of 77 K. To simulate the magnetic field distribution around the coil an air domain was used as an approximation to the pure nitrogen environment in the planned experimental assembly.

The superconducting domain's critical current was modelled using as input the measured $I_{\mathrm{c}}(B,\Theta)$ characteristic, as shown in Figure 5.2 [122]. The n-value was also available for various background fields and angles and was consequently applied in the model as $n(B,\Theta)$ (see Figure A.4 [122]).

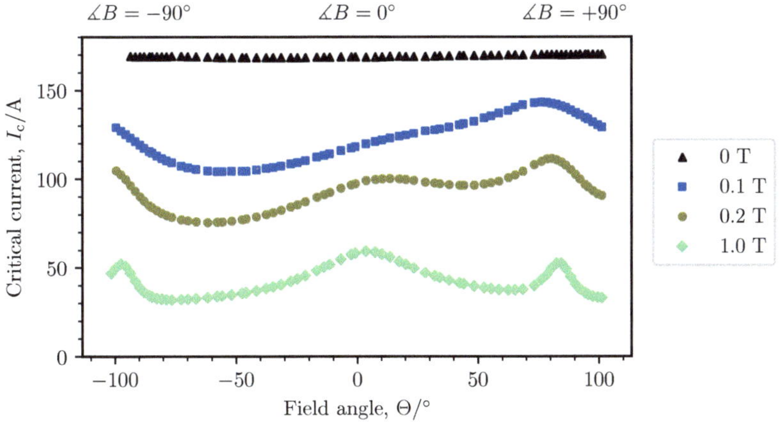

Figure 5.2: Dependence of the critical current on the orientation and amplitude of the applied magnetic field in a the 4 mm-wide SCS 4050AP HTS tape used in the coils at 77 K [122, 123].

5.2.1 Non-insulated coil

In the case of the NI coil, the current density may vary among the turns as current can bypass between them. Ignoring the resistive heating effects of such a bypass, a solution is sought where

$$J(x,y) = J_{\mathrm{c}}(x,y)(B,\Theta). \tag{5.5}$$

The critical current of the coil was then calculated by integrating the current density over all individual turns and finding the minimum

$$I_{\mathrm{c,\ min}} = \min \int_{n_{\mathrm{t}}} J_{\mathrm{c}}(B,\Theta)\mathrm{d}A_{\mathrm{t}}, \tag{5.6}$$

and mean

$$I_{\mathrm{c,\ mean}} = \mathrm{mean} \int_{n_{\mathrm{t}}} J_{\mathrm{c}}(B,\Theta)\mathrm{d}A_{\mathrm{t}}, \tag{5.7}$$

values. Here n_{t} represents a given turn number and A_{t} is the cross-section area of a tape. When applying $J_{\mathrm{c}}(B,\Theta)$ as driving current to the NI coil, the minimum value of the surface integral current was 52.7 A whereas the mean value was 73.1 A. These are 93 % and 109 % of the experimentally obtained critical current of 57 A and 67 A (using the 0.1 $\mu\mathrm{V\,cm^{-1}}$ and 1 $\mu\mathrm{V\,cm^{-1}}$ criterions), respectively (see Table 5.1).

5.2.2 Insulated coil

In the insulated coil additional boundary conditions were necessary to restrict the overall current in each turn to equality. In the first approach, the conductor was homogenized with the additional thickness of the Kapton® insulation[1], resulting in a total thickness of 210 μm. In the second approach, both the HTS tape and the insulation were modelled, with the electrical conductivity of the insulator layer set to zero. To restrict current flow between turns in the homogenized approach

$$\mathbf{n} \cdot \mathbf{J}|_{\Omega} = 0, \tag{5.8}$$

the boundaries were insulated electrically

$$\sigma|_{\Omega} = 0, \tag{5.9}$$

[1] DuPont™and Kapton® are trademarks or registered trademarks of E.I. du Pont de Nemours and Company.

where Ω denotes the boundary of the superconducting domains.

To find the critical current of the insulated coil with the given boundary conditions, a parametric sweep simulation was carried out, by applying uniform current from 0 A to 80 A to each turn of the coil. At each step the minimal critical current in the coil was calculated according to

$$I_{\text{coil}} = \int_{n_{\text{t}}} J_{\text{c}}(B, \Theta) \mathrm{d}A_{\text{t}}. \tag{5.10}$$

By plotting the calculated critical current as a function of the applied current, the critical current of the coil was obtained at the point that satisfies e.q 5.5.

Figure 5.3 shows the results of the simulation searching for the coil's critical current in steps of 1 A. The critical current line represents the limiting critical current of the coil, when supplied with a corresponding applied current, $I_{\text{c}}(B(I_{\text{app}}), \Theta)$. The intersection of the critical current and applied current line indicates the critical current of the coil as 74.5 A, representing a 3.5 % overshoot compared to the measured valued of 72 A (using the 1 µV cm^{-1} criterion).

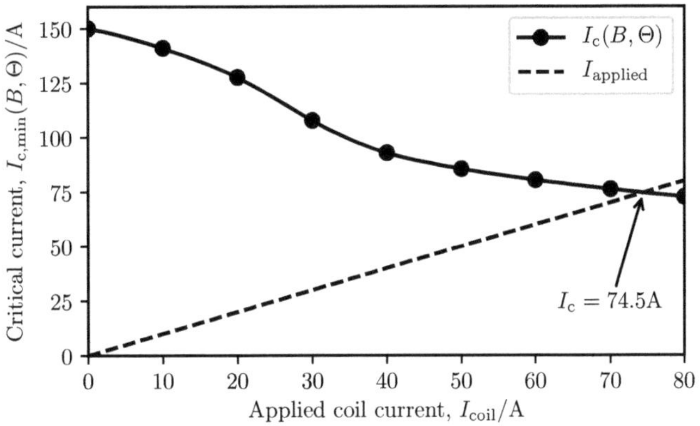

Figure 5.3: Critical current search results using FEM simulation in the insulated coil. The curve shows the critical current of the coil as a function of the magnetic field, generated by the applied current as shown by the dashed line. The intersection of the two curves indicates the critical current of the coil.

5.2.3 Coil parameters

The magnetic field distribution around the cross-section of the coils is shown in Figure 5.4. To compare this with measurement results, the calculated magnetic field strength at position $x, y = (0, 0.004)$ (location of the Arepoc Hall-effect sensor in the experiments) is plotted in the critical current measurements of the coils in Figure 5.15 and Figure 5.30 in Section 5.4 and Section 5.5. The calculated field strength of 250 mT to 260 mT at this point matches measurement results at currents corresponding to the simulated data, within error limits obtained in the I_c.

The stored energy of the coil e_{coil} in the FEM model was calculated as the integral of the magnetic field

$$e_{coil} = \frac{1}{2} \iiint \mathbf{B} \cdot \mathbf{H} dv. \tag{5.11}$$

From this then the coil inductance was obtained using

$$L = \frac{2e_{coil}}{I^2} \tag{5.12}$$

at the critical current of the coil, as calculated from the model.

A summary of the calculated values together with design parameters of the coil are presented in Table 5.1.

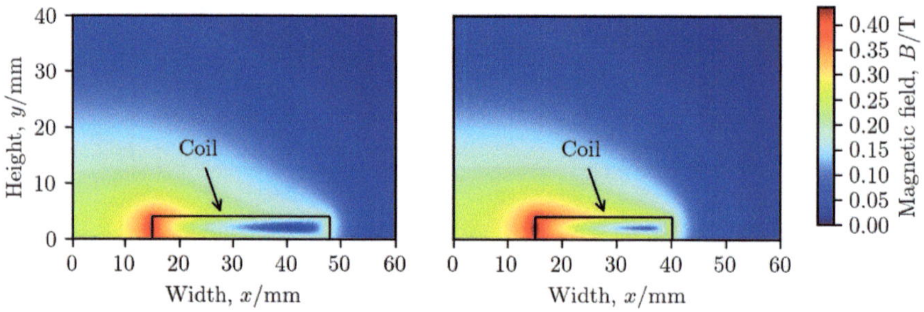

(a) Insulated coil (b) Non-insulated coil

Figure 5.4: 2D magnetic field distribution over the cross-section of (a) the NI-coil and (b) the insulated coil. The y-axes represent the axis of symmetry.

Table 5.1: Specifications and properties of the NI and insulated coils, wound from SuperPower SCS4050 AP tape. † At 77 K, self-field. [a] Using the $0.1\,\mu V\,cm^{-1}$ criterion. [b] Using the $1\,\mu V\,cm^{-1}$ criterion. [c] Calculated value. ‡ Represents the "ideal" field coefficient calculated from the maximum magnetic field and current.

Parameter	Symbol	Unit	Non-insulated	Insulated
Number of turns	N_t	-	157	157
Tape width	w_{tape}	mm	4	4
Inner and outer diameter	r_i, r_o	mm	30 and 85	30 and 97
Tape length	l_{tape}	m	28	31
Tape thickness	t_{tape}	µm	163	163
Coil mass	m_{coil}	g	334.78	270.08
Tape critical current †	$I_{c,\ tape}$	A	141	141
Coil critical current †, [a]	$I_{c,\ coil}$	A	57	65
Coil critical current †, [b]	$I_{c,\ coil}$	A	67	72
Coil critical current †, [c]	$I_{c,\ coil}$	A	52.7 and 73.1	74.5
Inductance	L	mH	1.38	1.49
Inductance [c]	L	mH	1.1	1.49
Discharge time constant	τ_{down}	ms	610	-
Field coefficient ‡	k	$mT\,A^{-1}$	3.98	3.6
Stored energy	e_{coil} (at I_c)	J	2.48	4.12

5.3 Coil preparation

This section discusses the main considerations in designing and building the experiment and winding the coils, focusing on the aspects of:

- Cryostat: cool the coil to 77 K, but avoid boiling liquid nitrogen.
- Design: suitable proof-of-concept coil while keeping its surface unobstructed.
- Coil winding: make sure the tapes are aligned and that sufficient tension is kept during winding. Co-winding with Kapton® to achieve an insulated coil.
- Electrical joints and instrumentation: properly solder electrical joints.
- Fluorescent coating: apply fluorescent coating to the coil's surface.

5.3.1 Cryostat

To avoid complex and expensive vacuum and cryocooler setups with optical windows – and since the aim was to test the coil only at 77 K – an in-house, 2-component cryostat was constructed. An aluminium container was placed inside a polystyrene container and the space between the two was filled up with liquid nitrogen, as shown in Figure 5.5. The coil was then placed in the aluminium box and a good thermal contact was made between the two with the help of a thin layer of thermal paste. This allowed the coil to be conduction-cooled to 77 K via the aluminium container's walls. The remainder space in the containers was filled with the evaporated, cold nitrogen gas, providing a cold and clear atmosphere, ideal for the optical measurement. The high-speed camera was then positioned above the setup and lowered close to the coil inside a separate optical cryostat that kept the camera at room temperature. The UV excitation light was also placed in the camera's cryostat, facing downwards. The setup was finally surrounded by multi-layer insulation from the top to prevent radiation heat loads, air movement and mixing with the cold nitrogen gas as well as to block out any ambient light that would interfere with the optical measurement.

The result was a conduction-cooled coil in a pure, cold, nitrogen atmosphere, where the continuous boil-off of liquid nitrogen provided a flow of cold gas keeping the atmospheric moisture out of the assembly. Cooling the coil from ambient temperature to 77 K was possible within 30 min as shown in Figure 5.6. However, to achieve complete thermal equilibrium in both the entire assembly as well as in the measurement devices, the coil was cooled for up to 3 h before the measurements started.

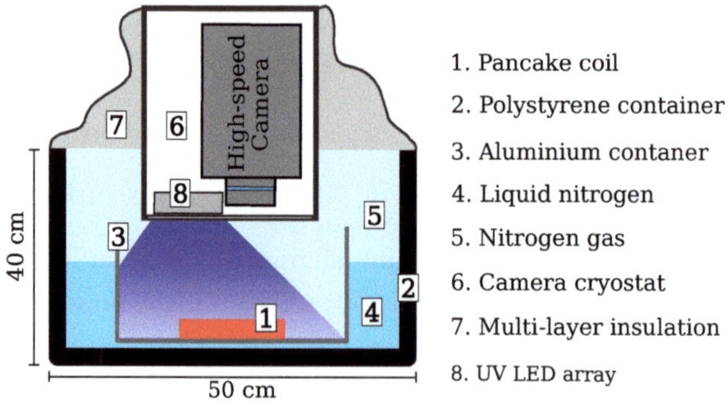

Figure 5.5: Cross-section drawing of the coil measurement assembly. The coil is fixed with thermal paste at the bottom of the aluminium container and excited by the UV LEDs from inside a separate cryostat housing the high-speed camera. Radiation heat load and ambient light are blocked by sheets of multi-layer insulation.

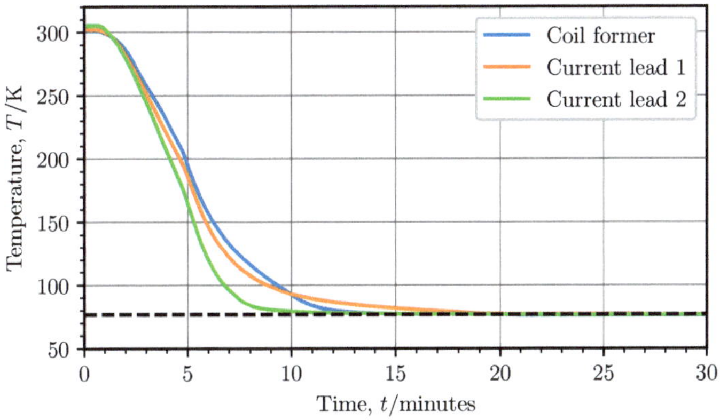

Figure 5.6: Temperature profile of the two current leads and the coil former when cooling the coil in the conduction-cooled assembly to 77 K. The dashed line indicates the liquid nitrogen temperature at atmospheric pressure.

5.3.2 Coil design

The coil design was governed by the need of a large, optically visible area during the measurement to be thermally imaged. Hence the inner coil former diameter was made as small as possible while the outer diameter as large as possible, keeping in mind the length of superconducting tape required and minimum bending diameter of the HTS tape [124]. As previously seen in Chapter 4 the boiling of liquid nitrogen severely impacts the recording quality and therefore it had to be avoided. As a consequence, the coil had to be conduction-cooled and the current leads had to be designed in a way not to obstruct the view of the coil.

Considering these aspects, a copper cylinder with a diameter of 30 mm and a height of 6 mm was prepared and served as the first current lead or contact. The cylinder was equipped with a slit from the outside until its centre line, where a 4 mm deep, M6 drill hole was present. To fix the HTS tape to the cylinder, the coated conductor was inserted into the slit in the cylinder and pushed into the centre hole. Here an M6 bolt with a vertical slit was pushed down onto the tape, which was then fixed by turning the bolt. This pulled in a portion of the tape and kept it mechanically stable for the first few turns of the winding. The bottom part of the same copper former had a 12 mm wide and ~0.5 mm deep groove where a 12 mm HTS tape was soldered to provide the first current lead. This is shown in Figure 5.7. The outer diameter of the coil was fixed at 85 mm, which required about 30 m of HTS tape and was constrained by availability.

At the bottom and around the copper former, a "c-shaped", 2 mm thick copper disk provided mechanical and cooling support for the coil. This copper disk had a 30 mm hole

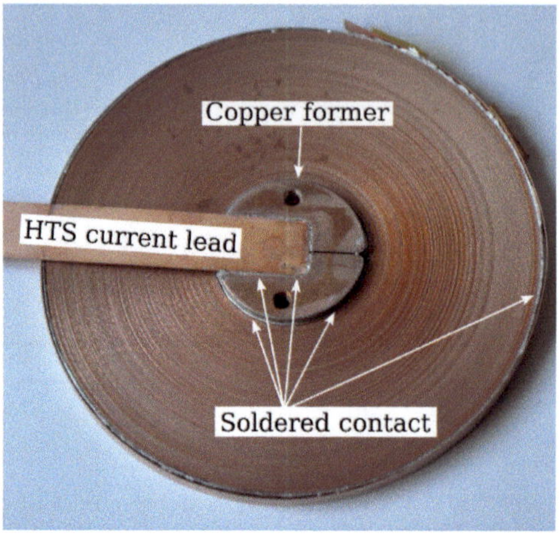

Figure 5.7: The back-side of the coil showing the 12 mm-wide HTS current lead as well as the coil's innermost windings soldered to the coil former.

at the centre – matching the diameter of the copper cylinder – as well as a 12 mm-wide slit in one direction to allow space for the HTS current lead. In the final assembly, the copper disk is insulated from the HTS current lead and the windings of the coil. The last two turns of the coil were soldered together, purely for mechanical reasons. The last turn of the winding was left to be ~20 cm longer to be used as the second current lead.

The current contacts were two copper blocks with a multi-contact plug attached to them using an M8 thread and insulated from the aluminium container via a sheet of Kapton® foil. These were clamping the HTS leads about 20 cm away from the coil using a thin sheet of indium between the HTS surface and the copper block. In the experiments it was found that by thermally anchoring the current leads in the liquid nitrogen bath, the temperature of the current contacts did not rise above 78.5 K when applying current up to 74 A.

The insulated coil was designed according to the same specification and principles as the NI coil, however, due to the additional thickness of the electrical insulation, the coil parameters for an insulated coil could not be kept the same as those of the NI coil. It was decided to keep the number of turns constant. As a consequence, the inductance of the coil as well as the peak magnetic field and limiting critical current are slightly different. Due to the added thickness of the Kapton® tape, the coil outer diameter was 97 mm compared to 85 mm for the NI coil.

The winding was done exactly the same as in the case of the NI coil, with the only difference that the turns were insulated from one another by a 4 mm-wide and 50 µm-thick Kapton® sheet. The instrumentation of the coil was done using the same voltage taps, Hall-sensor, the same current leads (including the Pt100 temperature sensors) and a new Pt100 temperature sensor for the centre of the coil, since this was bonded to the copper cylinder using X60, two-component adhesive.

5.3.3 Coil winding

For both coils, SuperPower, SCS4050 AP tape was used with a width of 4.1 mm and an effective thickness of about 163 µm, as governed by availability at our lab. Although the specifications of the tape state 40 µm copper stabilization on either side, the effective thickness of the copper layer is considerably thicker at some positions, as shown in Figure 5.8. Based on the dimensions of 30 mm inner diameter and 85 mm outer diameter, this resulted in a total of 157 windings.

The coil was wound using the copper cylinder as a former (or target) and a spool of the HTS tape on a source spool. Both were placed in level with one another to keep a consistent winding profile. To further keep the tape in position, a cylindrical nylon roll with a 4 mm groove around its circumference was used to guide the tape and smooth out any up/down swings resulting from the uneven alignment of the tape on the source spool. Pictures of the winding assembly illustrating the winding of the insulated coil are shown in Figure 5.9 and Figure 5.10.

Both the target coil and the source spool had a dc motor, the first for controlling the winding speed and direction and the latter for keeping tension in the tape by applying a constant torque. The dc stepper motor of the target coil was speed controlled by a LabVIEW software and the motor of the source spool was keeping tension in the tape, proportional to the applied current. The tape was aligned during the winding such that it was continuously, yet barely, touching the bottom copper disk and the result was a coil with a perfectly flat surface and a bottom flushed to the copper disk. This is shown in Figure 5.10a and the finished coil with an even surface in Figure 5.10b. Such an even surface was highly desired for an easier coating application.

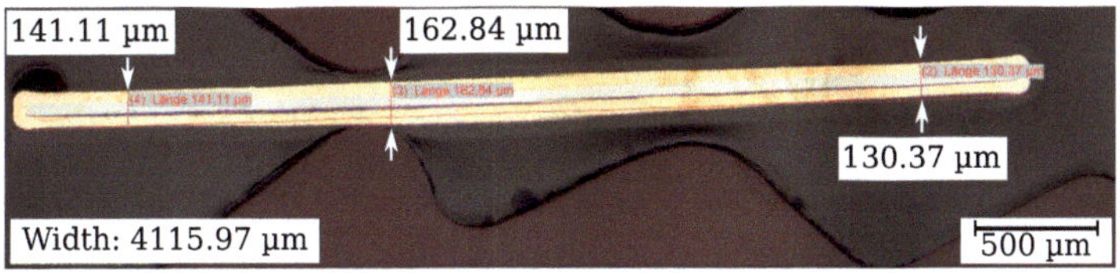

Figure 5.8: Cross-section microscopy image of a SuperPower SCS4050 AP tape used for winding the coils [125].

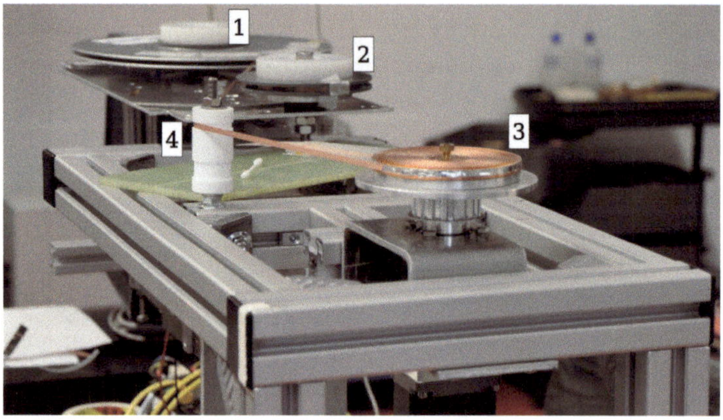

Figure 5.9: Picture of the winding machine assembly, highlighting: (1) the source spool of the 4 mm HTS tape, (2) the spool of Kapton® being co-wound with the HTS tape and (3) the target coil. Both tapes were passed through a nylon cylinder (4) to keep them aligned during winding.

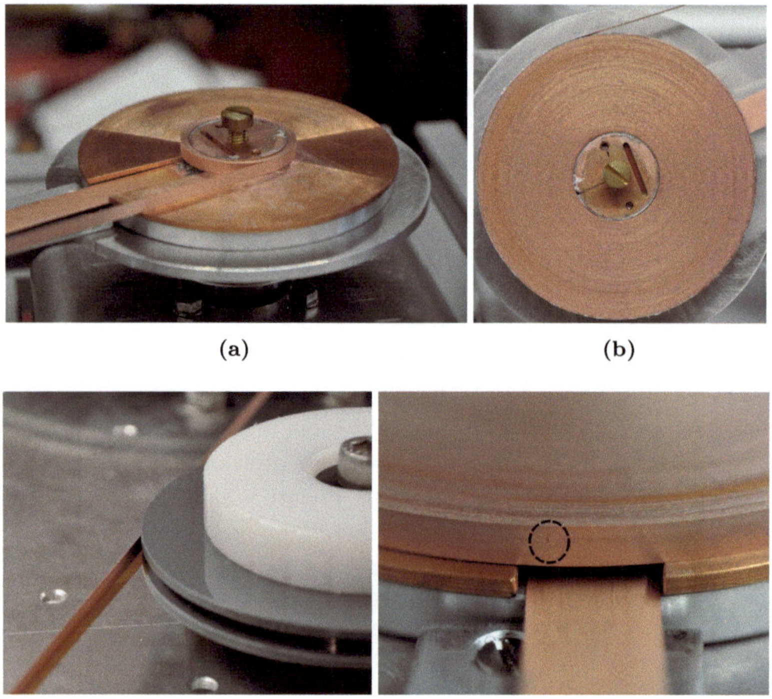

(a) **(b)**

(c) **(d)**

Figure 5.10: (a) A coil being wound onto the 30 mm former, showing the c-shaped cooling plate as well. (b) The finished coil on the winding setup with the soldered outer joint. (c) A close-up picture of the Kapton® co-winding with the HTS tape. (d) A small material defect on one of the insulated coil's outer windings. The point defect is later also visible in the finished coil.

Jo et al. reported the effects of winding tension and correlated I_c reduction on layer-wound coils [126]. They found that coils wound with 4.9 N to 78.4 N maintained constant ~90 % of the short sample I_c of the conductor. However, they went as far as 107.8 N, where considerable I_c reduction was already present. Although literature also suggests that higher winding tensions are possible, Kim et al. used 1 kg [74] and Yanagisawa et al. [127] 0.5 kg for winding a layer-wound coil and Jeon et al. [79] used 40 N. Initial experiments with 4 mm-wide pure copper tape indicated that higher tensions tend to strain the tape either on the top or along the bottom edge during winding. As a consequence, considering winding tensions reported in the literature and the need to avoid any possibility of damage, a relatively low winding tension of 2 kg to 2.5 kg was used.

5.3.4 Electrical joints

To provide good electrical contact and also to avoid high temperatures, a low melting point In-Ag (97 %-3 %) solder was used for all of the electrical joints. The melting temperature of this solder is 123 °C, which is below the temperatures that would permanently damage HTS coated conductors [128]. The soldering in all cases was done by first applying Spirflux 330® to the copper surface of the HTS tape or the copper former.

First, the copper cylinder was heated up and pre-tinned over the surfaces where the 12 mm HTS and the 4 mm HTS winding would later join. The 12 mm HTS current lead was also pre-tinned and soldered to the copper cylinder using a soldering iron. At this point the copper former – together with the soldered HTS current lead – was fixed on the winding setup. The 4 mm HTS tape was attached to the centre of the former and was pre-tinned over the length of the first winding. A 3 mm-wide In-Ag soldering tape was placed in between the copper former and the HTS coated conductor and the first few turns were wound to apply tension, similarly as shown in Figure 5.10a. The soldering at this point was done using a hot air blower until the soldering tape had visibly melted between the two surfaces.

The outer two turns were soldered in a similar way and the result is partially visible in Figure 5.9. Both the outer surface of the second last turn and the inner and outer surfaces of the last turn were pre-tinned. Again a 3 mm-wide In-Ag soldering strip was used between the turns, however in this case instead of a hot air blower a soldering iron was used. Going around the coil in the direction of the winding, the tape was heated until the solder melted between the outer two turns. This method allowed a good mechanical contact between the two last turns and avoided heating up the entire coil for soldering only a small section. Note that this also means that in the insulated coil, the last two turns are practically non-insulated as well.

The joint resistance was measured at current levels of 40 A and 50 A for both the insulated and non-insulated coils using a nanovoltmeter and was found to be approximately 200 nΩ and 600 nΩ, respectively. At these current levels the superconducting transition was not yet observable, but the higher current allowed measuring a voltage above the noise limit.

5.3.5 Coil instrumentation

A picture of the finished NI coil is shown in Figure 5.11 where the temperature sensors, voltage taps and the magnetic field sensor are shown. During measurements three temperatures were recorded: the two current leads, each with a Pt100 temperature sensor embedded in the copper blocks, and the central copper former. This latter had a 1.5 mm × 1.5 mm × 20 mm groove that allowed inserting a Pt100 temperature sensor. The sensors in the current leads were thermally bonded using silicone paste and fixed to the copper blocks by zip-ties. The sensor in the coil former was fixed in the groove using fast-drying, methyl methacrylate-based X60 glue. This served as both mechanical support as well as thermal bonding. For measuring the axial magnetic flux density B_z an Arepoc hall-effect sensor, with a 99.8 mV T^{-1} sensitivity, was fixed with double-sided duct tape at the centre of the copper cylinder, just above the M6 drill hole.

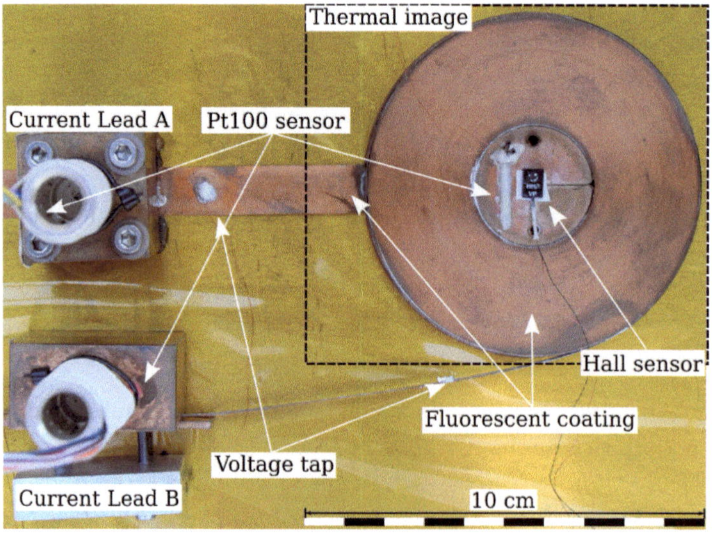

Figure 5.11: HTS non-insulated coil placed in the aluminium box and insulated from the bottom via Kapton® sheet. Three Pt100 temperature sensors, a pair of voltage taps, a Hall-effect magnetic field sensor as well as the area covered by the fluorescent coating are indicated by the white arrows. The black, dashed rectangle indicates the boundaries of the area recorded by the camera, in the same orientation as presented later in thermal images.

To avoid any distortion and additional unknowns in the coil, only a single pair of voltage taps were used, outside of the coil. These were soldered with In-Ag solder to the HTS current leads a few cm away from the coil. As a consequence, the voltage signal measured across the coil includes two resistive joints. First, between the 12 mm HTS current lead and the copper cylinder, and second, between the copper cylinder and the 4 mm windings of the coil. However, this resistance was measured to be insignificant for any practical purpose in this work.

The measurement devices used were closely identical to that described in Chapter 4 with the difference that for the coil several additional data acquisition channels were used, including a parallel-connected nanovoltmeter measuring the voltage drop across the coil. The Dewetron amplifier was used for four channels: current across an electrical shunt resistor, voltage drop across the coil, camera trigger signal reference and the magnetic field from the Hall sensor. In addition, three Dataforth SCM5B40/41 amplifier modules were used to acquire data from the Pt100 temperature sensors. All channels, except for the nanovoltmeter, were acquired using the NI USB 6289 DAQ.

In the case of the NI coil it was found that the Heizinger current source, used also for the pulsed HTS tape measurements, was accurate enough to carry out the experiments, despite the presence of some current ripples. These were mostly "filtered" by the RL characteristics of the coil. In the case of the insulated coil, however, even the smallest fluctuations in current caused large inductive voltage responses and hence made accurate measurements impossible. Therefore a more accurate, Cryogenic SMS120C, magnet ramping source was used that was specifically designed for inductive loads. The available ramping rates were slightly different for this power supply, with the maximum being $12\,\mathrm{A\,s^{-1}}$.

The insulated coil – after measured for various ramping rates – was retrofitted with a $5\,\Omega$ resistor (at 77 K) to be used as a local heater and test the coil's stability, shown in Figure 5.12. The same X60 glue was used as for the Pt100 temperature sensor at the centre of the coil and it was positioned approximately 1 cm away from the coil former, opposite to the side of the HTS current lead. The heater was deliberately positioned in the "body" of the coil, to allow the observation of heat flux, not necessarily at the centre of the coil.

5.3.6 Fluorescent coating

To observe local heating, the coil's surface was covered with the EuTFC-based fluorescent coating. Several tries were made including an airbrush and using a traditional paint brush to apply the coating to the coil's surface, however the result in all cases was not satisfactory as the coating became either too thin or highly non-uniform. For example,

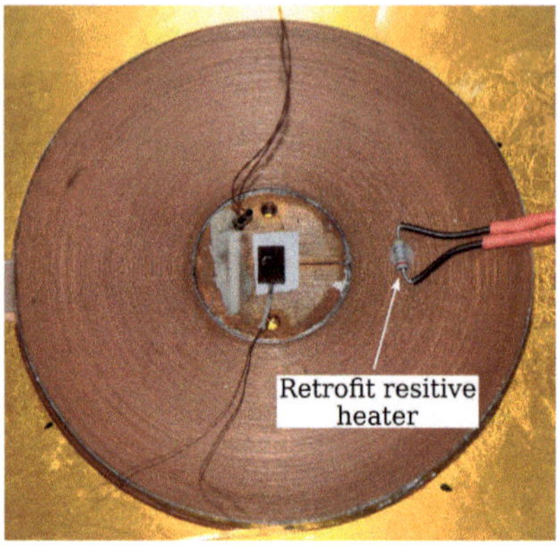

Figure 5.12: A 5 Ω resistor retrofitted and used as a heater on the surface of the insulated coil.

using an airbrush proved to be difficult twofold: first, the coating has no colour and therefore it was difficult to see which part was coated already and how uniformly. Second, applying a successive layer over a coated area to improve the coating thickness is not feasible, as the acetone from the coating would dissolve any already existing coating. The only solution seemed to be the use of a single layer coating rapidly sprayed over the entire surface. However, due to the highly diluted coating solution, this method was not applicable and yielded poor results.

It was found that the same droplet coating method, as used when measuring single HTS tapes as described in Chapter 4 [129], worked best for the HTS coils as well and provided a sufficiently thick and uniform coating for the measurements. Figure 5.13 shows the quality of the final coating, where bright parts indicate either a higher optical reflectivity (e.g. copper surface) or a thicker fluorescent coating, whereas darker parts indicate a thinner fluorescent layer. Most of these optical artefacts were cancelled out during post-processing.

The initial concern with this method – applied to a pancake coil – was the possibility of the acetone-PMMA solution entering the space between the windings and forming an insulation at certain positions. However, in lack of a better approach for applying the coating and given the tight coil winding it was still adopted and no related problems were encountered later on. Nevertheless the coating was significantly more difficult than the coating of single HTS tapes. In the case of the NI coil, the coating did remain mostly on the surface, even if some of it entered in between the turns. With the insulated coil, however, problems were encountered due to the slight height difference between the HTS tape and the Kapton® co-winding. These two have effectively created a spiral-shaped groove on the surface of the coil. When applying the liquid coating these grooves hindered the distribution and caused highly non-uniform light intensity over

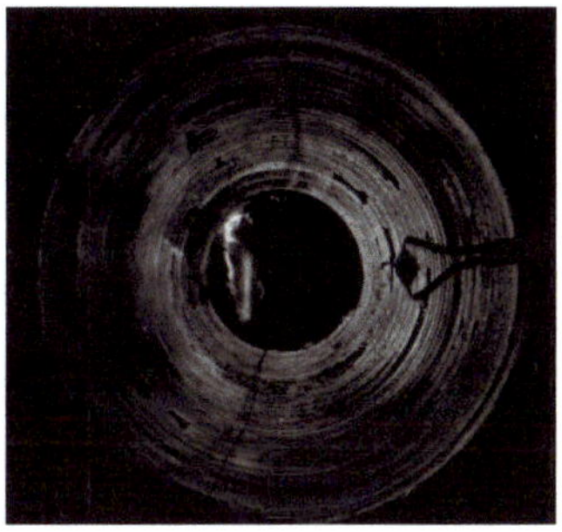

(a) Insulated coil

(b) Non-insulated coil

Figure 5.13: Raw fluorescent image of (a) the insulated coil and (b) the NI coil. The image of the insulated coil already shows a retrofit resistor used as a heater.

the surface as shown in Figure 5.14. Increasing the amount of coating applied to the surface did not help as described above on the example of the air brush and at the same time caused more coating to disappear in between the turns. For a satisfactory coating it was found that instead of applying precise droplets continuously over the coil (inherently resulting in non-uniform coating) applying the coating from a larger height of 10 cm to 20 cm was more appropriate. This allowed the droplets to splash in a larger radius and self-distribute better over the surface while avoiding excessive coating solution build-up.

The high-speed camera, despite used at several thousand fps to measure quench in single HTS tapes (refer to Chapter 4, [129]), was used at the slowest speed setting of 50 fps, which was found sufficient for measuring the heating in the coils. Depending on the position of the coil and the camera, this allowed an approximate recording time of about 10 min to 12 min per experiment. In the post-processing stage the recordings

(a)

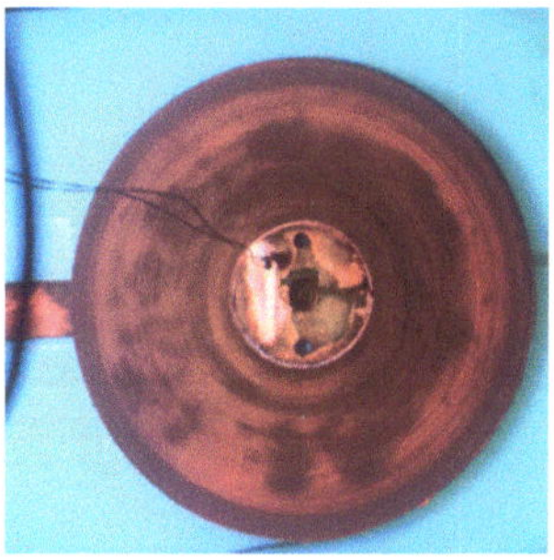

(b)

Figure 5.14: (a) shows the NI coil's surface completely covered with the coating solution. In (b) the fluorescent emission from the coating shows a strong non-uniformity and partially missing spots.

were generally averaged over 200 ms to reduce noise, without any loss of information about the transient effects.

5.4 Non-insulated coil measurement results

5.4.1 Critical current

The coil was initially measured in liquid nitrogen to safely determine the operating margins. Figure 5.15 shows the results of this I_c measurement, where using the $0.1\,\mu V\,cm^{-1}$ and $1\,\mu V\,cm^{-1}$ criterion the critical currents were determined as 57 A and 67 A, respectively. The magnetic field development, as simulated by the FEM model, shows a lower expected field than experimentally measured. This is due to the difference in current densities in the model and experiment as shown on the calculated I_c values in Table 5.1. After concluding all experiments presented in this section, another I_c measurement was performed to check whether the coil was damaged in any way. It was found that the critical current of the coil did not degrade in any measurable quantity.

5.4.2 Turn-to-turn resistance

The importance of the turn-to-turn contact resistivity in NI coils was introduced previously in Chapter 2. By knowing the winding tension, R_c can be calculated from eq. 2.17 provided that measured data is available for R_{ct}. In this case one must assume

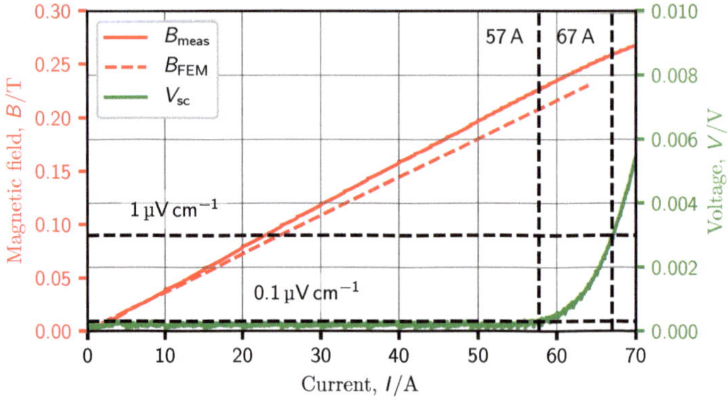

Figure 5.15: Initial I_c measurement of the NI coil showing the $0.1\,\mu V\,cm^{-1}$ and $1\,\mu V\,cm^{-1}$ electric field criterion. The critical current of the coil is then 57 A and 67 A, respectively. The plot shows data points extracted from the measurement and not the actual ramping plot.

that the initial measurements are accurate enough as well as that the winding tension is constant during winding. For measuring the turn-to-turn contact resistivity, the same setup was used as by Otten [130]. Two pieces of HTS tape were placed face-to-back on top of one another with an overlap of 4 cm, and were put in a hydraulic press and cooled to 77 K in liquid nitrogen. The hydraulic press was operated manually by applying increasing pressure on the tape and, simultaneously, a 10 A dc current was applied across the overlap of the two tape pieces. The pressure dependence of the resistance was obtained by measuring the voltage drop across the interface between the two HTS tapes as well as the pressure by the use of strain gauges. To accurately reproduce the condition of the demonstrator coils, the samples were taken directly from the spool and were not cleaned before measurement.

The measurement results in Figure 5.16 show a surface contact resistivity of around $300\,\mu\Omega\,\mathrm{cm}^2$, which lies in the range reported by Otten [130] in a more extensive measurement using the same face-to-back configuration. When substituting this value into eq. 2.17 to calculate the coil resistance numerically

$$R_\mathrm{c} = \sum_{i=1}^{156} \frac{3 \times 10^{-8}\,\Omega\,\mathrm{cm}^2}{2 \cdot \pi \cdot (0.0015\,\mathrm{m} + i \cdot 163 \times 10^{-6}\,\mathrm{m}) \cdot 0.004\,\mathrm{m}} = 6.6\,\mathrm{m}\Omega. \tag{5.13}$$

However, deriving the characteristic coil resistance using eq. 2.9 with the discharge time constant and inductance of the coil resulted in

$$R_\mathrm{c} = \frac{1.38 \times 10^{-3}\,\mathrm{H}}{610 \times 10^{-3}\,\mathrm{s}} = 2.23\,\mathrm{m}\Omega. \tag{5.14}$$

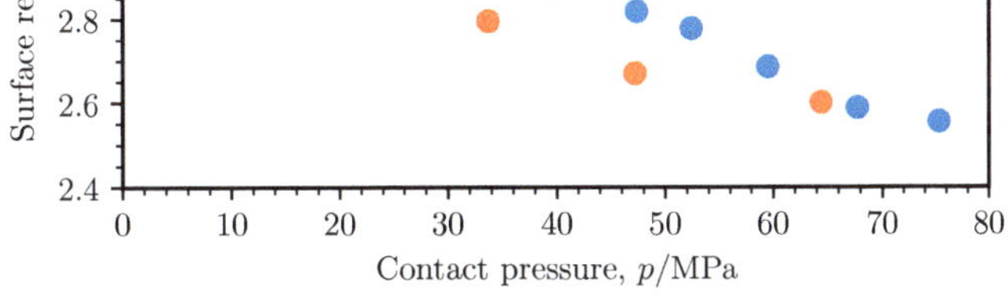

Figure 5.16: Surface contact resistivity measurement at 77 K on two pieces of SCS4050 AP tape from SuperPower during a single load cycle.

The two equations show a three-fold difference in the coil resistance when calculated using the two different approaches. This suggests either different surface characteristics between the measured tape sample and the long length HTS used in the coil, measurement inaccuracies or some fundamentally different behaviour.

5.4.3 Thermal and electrical effects of various ramping speeds

In the conduction-cooled experiments, the coil was ramped using several ramping rates, while being recorded at 50 fps to observe whether the different ramping rates caused any heating. The tested rates were $0.1\,\mathrm{A\,s^{-1}}$, $0.5\,\mathrm{A\,s^{-1}}$, $1\,\mathrm{A\,s^{-1}}$, $2\,\mathrm{A\,s^{-1}}$, $5\,\mathrm{A\,s^{-1}}$, $10\,\mathrm{A\,s^{-1}}$, $20\,\mathrm{A\,s^{-1}}$, $50\,\mathrm{A\,s^{-1}}$ and $100\,\mathrm{A\,s^{-1}}$ with the target current levels being 40 A, 50 A and 55 A. A representative part of the current and magnetic field developments are shown in Figure 5.17 during coil ramping.

The electrical results of a single, typical ramping measurement are shown in more detail in Figure 5.18 where the coil was ramped with $10\,\mathrm{A\,s^{-1}}$ to 55 A[2]. In each measurement the coil was ramped with a specific speed and upon reaching the target current, it was given 60 s to reach equilibrium. This was then followed by a ramp down, with the same speed, followed again by a 60 s final relaxation time. In order to allow the coil to return into thermal equilibrium after each measurement, a ~5 min beak was kept between any two experiments and the liquid nitrogen level was continuously monitored.

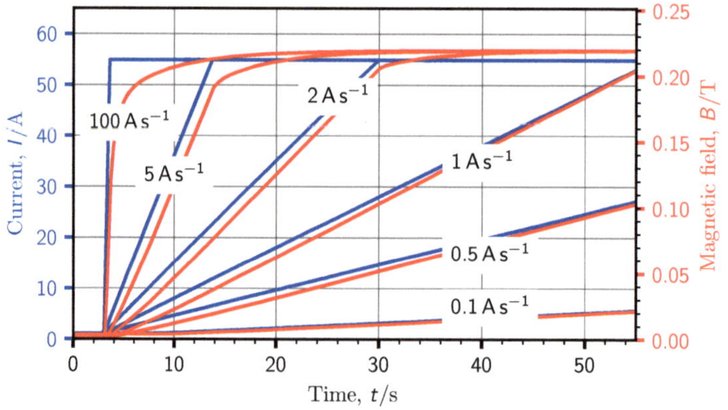

Figure 5.17: Time development of current and magnetic field in the non-insulated coil during the first 60 s when ramped at various ramping rates to 55 A.

[2] For simpler visual representation, the inductive voltages from the measurements are always represented as positive in the figures.

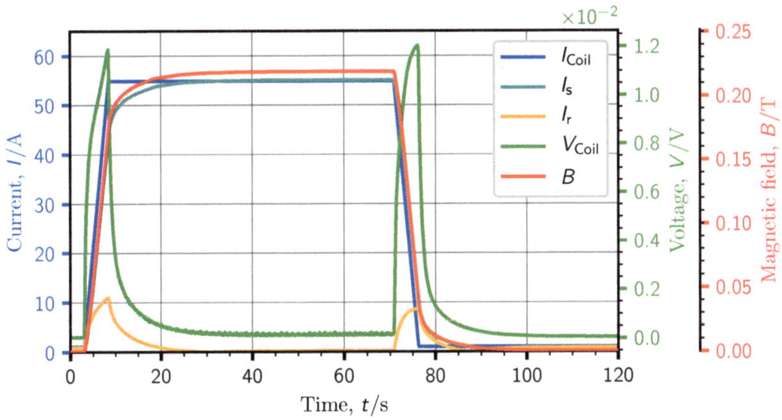

Figure 5.18: Time development of measured current, voltage and magnetic field in the non-insulated coil when ramped at 10 A s^{-1} to 55 A. The radial and the spiral current (I_r, I_s) were calculated using eq. 5.15.

Due to the inductance and lack of turn-to-turn insulation of NI coils, during a current ramp with an arbitrary dI/dt a voltage is induced according to eq. 2.11. While normally this would act against current flowing in the spiral path, in an NI coil the current may also bypass the turns across the turn-to-turn resistances. Hence the higher the inductance and the ramping rate, the more current will initially bypass the turns – or the coil itself – and revert back into the spiral path with a delay. The inverse happens in a sudden discharge, where the current supply is suddenly disconnected from the coil, e.g. by means of a switch (Figure 2.10), and as a consequence the coil sees a large dI/dt and develops a large voltage. This voltage then forces currents to flow through both the superconducting spiral path as well as the turn-to-turn resistances and dissipate quickly.

Assuming that it is predominantly the current in the spiral path I_s that generates the magnetic field, the current in the radial path I_r may be approximated as

$$I_r = I_{coil} - \frac{B}{k}. \tag{5.15}$$

Here k is the coil field constant defined as

$$k = \frac{B(t)}{I(t)}. \tag{5.16}$$

Since for an NI coil the magnetic field has a delay, k was obtained experimentally after the magnetic field has stabilized as

$$k = \frac{B_{\mathrm{max}}}{I_{\mathrm{max}}} = 0.003\,98\,\mathrm{T\,A^{-1}}. \tag{5.17}$$

Knowing the amount of current bypassing through the turn-to-turn resistances, the heating P and absorbed energy Q may also be calculated using

$$\begin{aligned} Q_{\mathrm{coil}} &= \int_0^{\infty} P \, \mathrm{d}t \\ &= \int_0^{\infty} I_{\mathrm{r}}^2 R_{\mathrm{c}} \, \mathrm{d}t \\ &= \int_0^{\infty} \left(I_{\mathrm{coil}} - \frac{B}{k} \right)^2 R_{\mathrm{c}} \, \mathrm{d}t. \end{aligned} \tag{5.18}$$

This equation, however, may only apply to the ramping and steady-state operation, when the coil is still in its superconducting state. After the superconductor transition begins, the energy stored in the magnetic field of the coil will also begin dissipating and it is more appropriate to use

$$\begin{aligned} Q_{\mathrm{coil}} &= \int_0^{\infty} P \, \mathrm{d}t \\ &= \int_0^{\infty} I_{\mathrm{coil}} V_{\mathrm{coil}} \, \mathrm{d}t. \end{aligned} \tag{5.19}$$

However, calculating the heat load using eq. 5.18 may result in uncertainties due to the uncertainty in the value of R_{c} itself as pointed out in Section 5.4.2. While several groups have proposed complex equivalent circuit grid models to simulate heating in NI coils [86, 117], the development of such models is beyond the aims set forth in this experimental investigation. Instead, to approximate the current sharing between the turns, the charging profiles were compared with the coil field constant and the peak current bypass value was calculated from eq. 5.15. The results are shown in Figure 5.19 for all ramping measurements. The plot shows that – as expected – a faster ramp causes higher (peak) radial bypass currents, with less than 5 A of bypass until a charging rate of $2\,\mathrm{A\,s^{-1}}$.

In three of the (time-wise consecutive) measurements the coil's lower-right sections showed signs of localized heating, later developing into a semi-circle and disappearing during ramping down. In these three experiments the coil was ramped with $50\,\mathrm{A\,s^{-1}}$, $100\,\mathrm{A\,s^{-1}}$ and $0.1\,\mathrm{A\,s^{-1}}$ to 40 A, 40 A and 50 A, respectively. The measurements are shown in Figure 5.20-Figure 5.22 with selected time excerpts of the thermal imaging along with the electrical measurements. As visible, most of the coil remained at close to

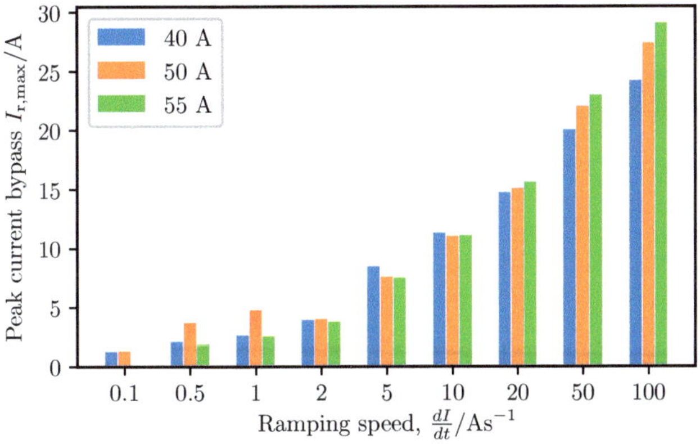

Figure 5.19: Estimated current bypass across the turn-to-turn contact resistances, calculated using eq. 5.15, at various current ramping speeds.

77 K, with the exception of a select area. In Figure 5.20b and Figure 5.20c the spreading of the warmer zone is visible just after the end of the ramp and approximately half way through the 60 s plateau. This warmer zone then dissipated when the coil discharging began. Similar effects are visible in Figure 5.21b and Figure 5.21c showing the 100 A s^{-1} ramping measurement, where heating is still visible around the same zone as in the previous measurement, although to a lower extent.

The effects are more expressed in the measurement ramping to 50 A as shown in the thermal images Figure 5.22b and Figure 5.22c. Here, apart from the distorted ring-shaped heating in the coil, it would seem that the heating is concentrated to the path directly between the inner and outer current contact, despite the slowest ramping rate tested.

Generally, in some of the measurements with target currents of 40 A and 50 A the coil's innermost windings showed moderate heating (80 K to 85 K) around the centre copper former. However, the same heating was not observed when targeting 55 A. In this latter case, some overall heating was observed in the coil body, yet to very moderate levels of ~80 K and without any deductible patterns with regards to ramping speed.

As a summary, in select measurements different parts of the coil were observed to warm up to different levels, however no clear pattern was recognized with respect to ramping rate or target current level (below the coil's I_c). Therefore, from the thermal imaging results, it cannot be clearly stated whether faster ramping rates would incur in increased heat load in an NI coil. However, it has to be kept in mind that the investigated coil's inductance and characteristic resistance are low and may be the cause of this result. While approximately a 5 min recovery time was kept between all ramping measurements to allow the coil returning into a steady state, it is possible

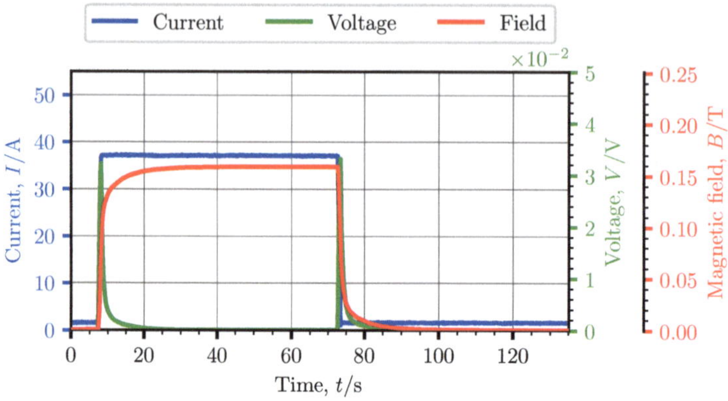

(a)

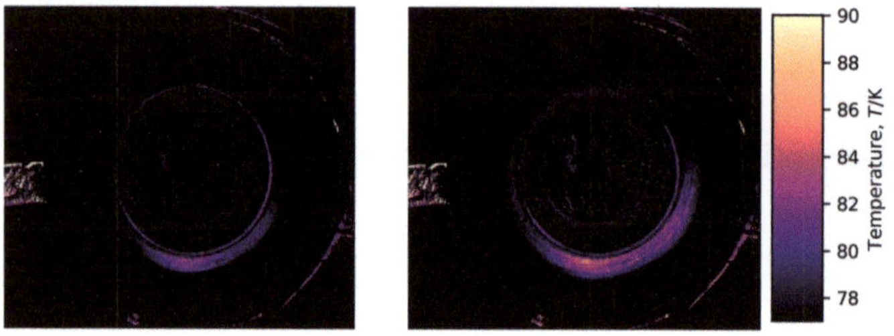

(b) $t = 19$ s

(c) $t = 34$ s

Figure 5.20: Electrical and thermal effects of $50\,\mathrm{A\,s^{-1}}$ ramping rate to 40 A.

that this time window was not sufficient, which may also induced anomalies into the measurement.

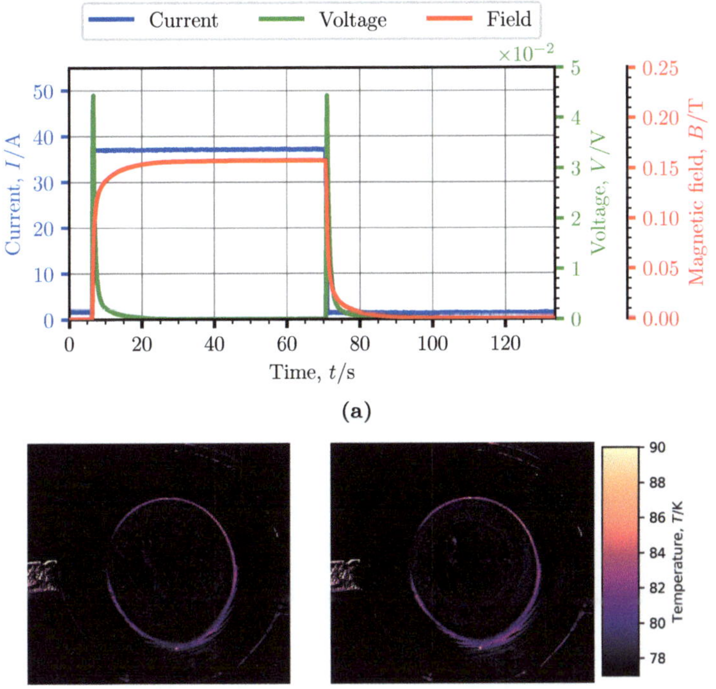

(a)

(b) $t = 19$ s

(c) $t = 34$ s

Figure 5.21: Electrical and thermal effects of $100\,\mathrm{A\,s^{-1}}$ ramping rate to $40\,\mathrm{A}$.

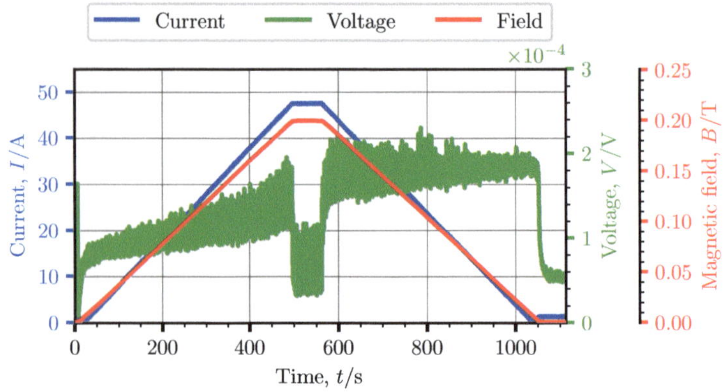

(a)

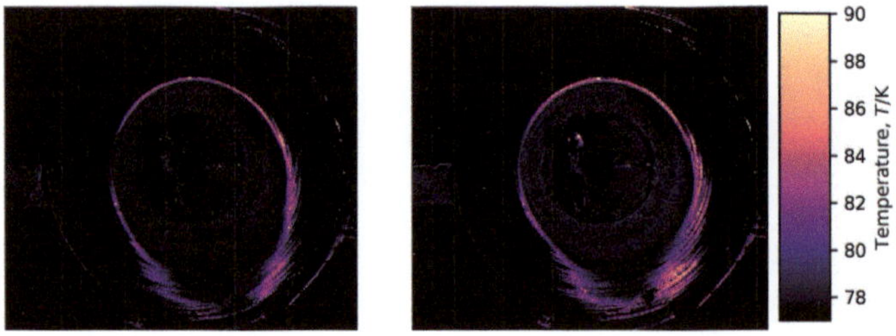

(b) $t = 32$ s

(c) $t = 60$ s

Figure 5.22: Electrical and thermal effects of $0.1\,\mathrm{A\,s^{-1}}$ ramping rate to 50 A.

5.4.4 Charging delay and charging time constant

Figure 5.23 shows the summary of all experimentally obtained charging delays of the coil when ramped to 40 A, 50 A and 55 A with ramping rates ranging from $0.1\,\mathrm{A\,s^{-1}}$ to $100\,\mathrm{A\,s^{-1}}$. The trend of the charging delays is clearly visible, and as expected [75], a faster ramping rate resulted in a longer time delay. The $0.1\,\mathrm{A\,s^{-1}}$ measurement shows negative values, however, this is merely an artefact from the calculation. At such a slow ramping rate the field increases closely with the coil field coefficient and hence at the time when the ramping ends, the 99.3 % field criterion is already reached. During data analysis this then may result in negative values. Interesting is also that with higher target current, the charging delay was moderately, yet consistently shorter. This can be caused by the field saturation of the coil, where more current simply does no longer cause a linear rise in the field and subsequently causes a faster perceived settling time. The plot also shows a saturation effect of the charging delay, where the difference between $10\,\mathrm{A\,s^{-1}}$ and $100\,\mathrm{A\,s^{-1}}$ charging only causes a moderate few second difference.

In Figure 5.24 it is shown that the coil charging time constant (as introduced in Chapter 2, [75]) is in fact a function of the ramping speed as well, where the higher the ramping rate, the shorter the time constant. This is supported by the fact that a faster ramping rate causes a higher voltage according to eq. 2.11, which in turn causes more current to flow in the radial direction, through a larger number of turn-to-turn contact resistances. This effectively increases the coil characteristic resistance and hence results in a lower time constant in eq. 2.9. Interestingly, Kim et al. [75] observed a correlation between the charging time constant and the target current level. By using

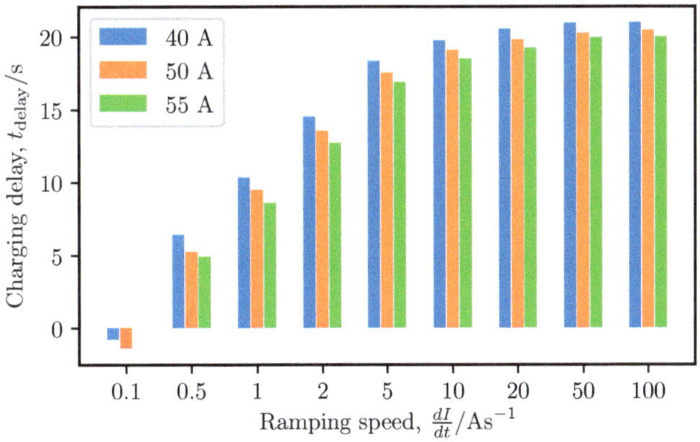

Figure 5.23: Data for the $0.1\,\mathrm{A\,s^{-1}}$ ramping rate to 55 A is missing.

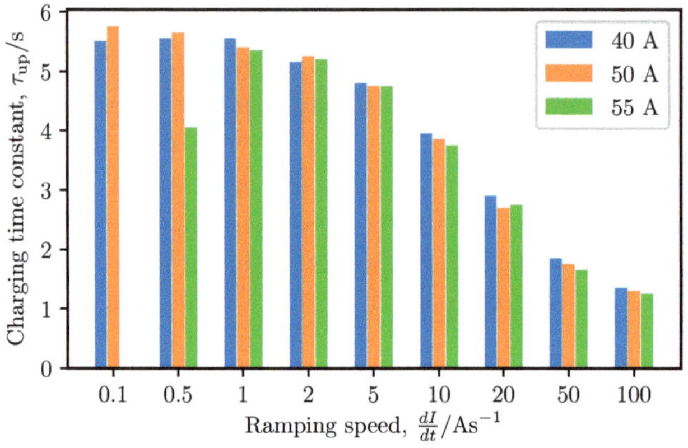

Figure 5.24: Charging time constant τ_{up} of the NI coil as a function of ramping rates between 0.1 A s^{-1} to 100 A s^{-1}. Data for the 0.1 A s^{-1} ramping rate to 55 A is missing.

a constant ramping rate in a racetrack pancake coil, they found that ramping to higher current levels caused a longer time constant. This was not observed in this work.

5.4.5 Sudden discharge

The coil discharge time constant was previously introduced in Section 2.4.2. To determine this experimentally, as well as to determine if any local heating is present in an accidental discharge, the coil was ramped to 30 A and 40 A in two successive measurements, where it was held for 60 s to allow for the centre magnetic field to stabilize. After this, the switch shown in Figure 2.10 was opened, allowing the coil to rapidly dissipate the stored magnetic energy in the coil resistance. The measurement results in Figure 5.25 show that the discharge time does not depend on the current amplitude and it is approximately 610 ms. This observation matches that of Wang et al. [76], where the same independence of the discharge time constant from the current amplitude was observed. Additionally, so have Kim et al. [75] come to the same conclusion with current amplitudes of 20 A, 40 A, 60 A and 80 A.

The energy stored in a coil at a given current level was calculated as

$$e_{\text{coil}} = \frac{1}{2}LI^2. \tag{5.20}$$

By knowing the discharge time constant the power dissipation was then calculated as

$$P_{\text{discharge}} = \frac{e_{\text{coil}}}{\tau_{\text{down}}}. \tag{5.21}$$

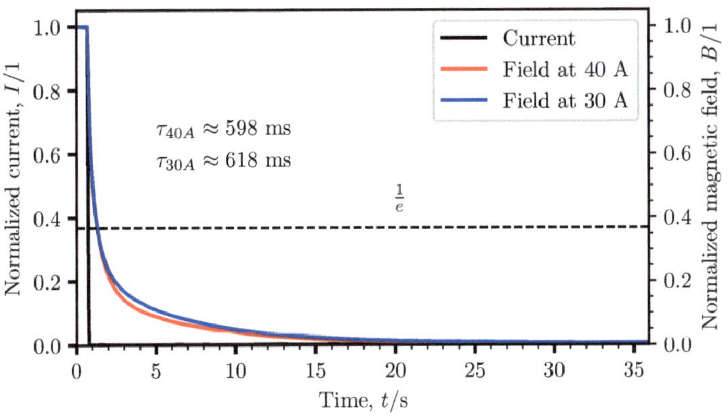

Figure 5.25: Normalized magnetic field development during sudden discharge tests at steady DC current levels of 30 A and 40 A. The dashed line indicates the 0.37 criterion where the time constant is evaluated.

From this the energy stored in the coil when held at 40 A was 1.1 J and the power dissipation approximately 1.8 W for a duration of 610 ms. With a coil mass of ~0.270 kg and by approximating the heat capacity with that of pure copper at 77 K as 195 J kg^{-1} K^{-1} [52, 131], the overall heat capacity of the coil would be 52.65 J K^{-1}. Disregarding heat distribution and the effects of cooling, it is clear that the energy released during a sudden discharge was not enough to cause any heating in the coil. This is merely a consequence of the small size and inductance of the test coil. In a larger coil with significantly larger inductance local heating may occur and make additional protection measures necessary.

5.4.6 Overcurrent

Measurements were done above the limiting critical current of the coil to test its behaviour without any artificial defect as to observe how a "natural" thermal runaway would happen. In overcurrent measurements, the coil was first ramped to a previously validated safe current level and from there slowly to higher currents. Every few amperes the ramping was stopped and the coil held steady to observe if the coil voltage was still stable. This stepwise ramping was then continued until the coil voltage would no longer stabilise and a "natural" runaway would start to appear. Special care was taken to manually ramp down the coil in case of a high voltage to permit future experimenting and avoid any damages. In different measurements, the transition point was observed at slightly different current levels, which suggests that slight disturbances in cooling, ramping and current levels can influence the thermal stability of the NI coil.

Overcurrent at 72 A

In one of the experiments the coil was initially ramped to 71 A at a rate of 10 A s^{-1} and the voltage was monitored, as shown in Figure 5.26. This eventually stabilized and so the coil was ramped higher to 72 A, this time at a slower 1 A s^{-1} rate, as shown at time 160 s. At this current level a slow, yet steady voltage shift was observed and so the coil was held at this level.

Since the magnetic field also reduced with the increasing voltage, the current must have been changing from the spiral to the radial path. This is also supported by the quasi-linear voltage rise, as expected with increasingly more current flowing through a resistor. In Figure 5.26b it is visible that initially the outer turns of the coil were at a higher temperature, presumably due to slightly worse cooling. At the same time, distinct heating is visible in the form of circles with varying width (Figure 5.26c and Figure 5.26d, as shown by the white arrows), indicating that over some turns the current was bypassing across the turn-to-turn contacts and caused losses. However, at about 298 s into the measurement, the voltage dropped and the magnetic field increased sharply, indicating that the current reverted — for an unknown reason – back into the spiral path quite abruptly. Simultaneously, the centre windings of the coil heated up, as shown in Figure 5.26e, where the ring-like hot areas were replaced by a uniformly warmer zone, mostly on the left-hand side of the coil. This showed that current flowing though the turn-to-turn contacts of the coil can be seen on the thermal image as heating over possibly several turns, which is most likely a function of the cooling performance and the distribution of local defects in J_c along the length of the wound conductor.

Overcurrent at 74 A

To reproduce previous results and observe the same behaviour again, the coil was ramped in a subsequent measurement directly to 72 A at a rate of 10 A s^{-1} as shown in Figure 5.27. However, in this case no voltage drift was observed. The coil was therefore ramped to 73 A and finally 74 A at a slower 1 A s^{-1} rate, where the coil voltage became unstable and started increasing once again.

Over the course of two minutes at steady 74 A, the coil voltage was rising continuously up until its final value of 0.47 V, where the coil was manually ramped down quickly to prevent any damage to the superconductor windings. During the transition, the magnetic field started decaying with the developing voltage, indicating a slow current transfer into the radial path. Then after 337 s a much sharper voltage rise and accompanied field drop is visible, likely due to (some of) the HTS tapes becoming resistive.

In Figure 5.27b it is visible that the outside of the coil is again at a higher temperature, however still cryostable. Warmer zones are still visible around the centre and middle layer of the coil, indicating heating due to currents bypassing across the turn-to-turn contacts. Figure 5.27c shows that the middle layer seems to have cooled back mostly and at the same time, due to the voltage rise of the coil, the innermost windings started heating up. At the bottom of the image, where the outer current lead arrives at the coil, a hot spot is also slightly visible. Then in Figure 5.27d distinct heating is visible around the centre coil former. Finally in Figure 5.27e, just before the coil was ramped down to protect it, the hottest temperature of around 100 K was reached at the few innermost windings of the coil. Note that the heating is limited to the innermost perhaps 1 cm of windings and that the outer turns are virtually unaffected. It is also interesting to observe that at the time when the coil voltage started rising, the middle turns of the coil actually cooled down and when the voltage continued to rise, the central windings started heating up instead. Most likely when the coil voltage is low, only the "major" defects in the coil would be bypassed, however at higher voltages more current could be forced across all turn-to-turn contacts, thereby reducing the current in the superconductor as well as local heating. Due to the larger overall thermal mass this could then manifest in a form of a local cooling effect. At the same time, the centre windings see the combined effect of superconducting to resistive transition of the HTS due to the reduced temperature margin and high magnetic field, combined with the current bypass and Joule heating due to the resistive joints.

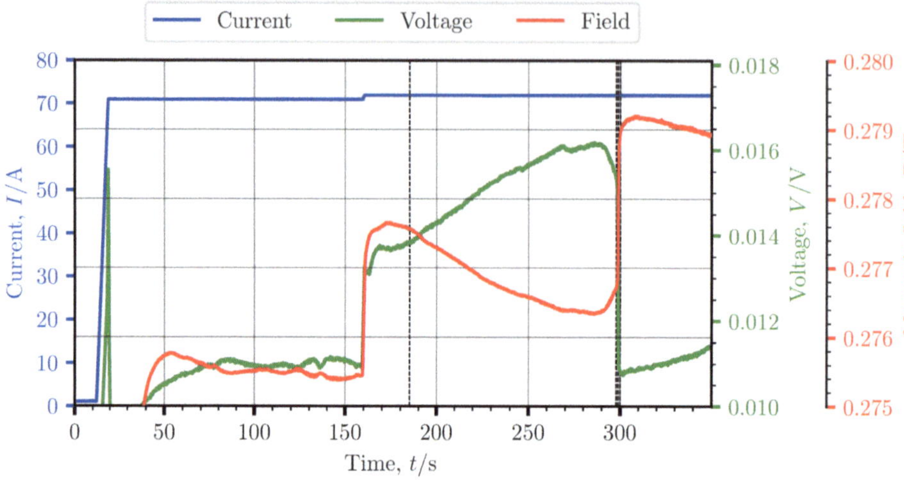

(a)

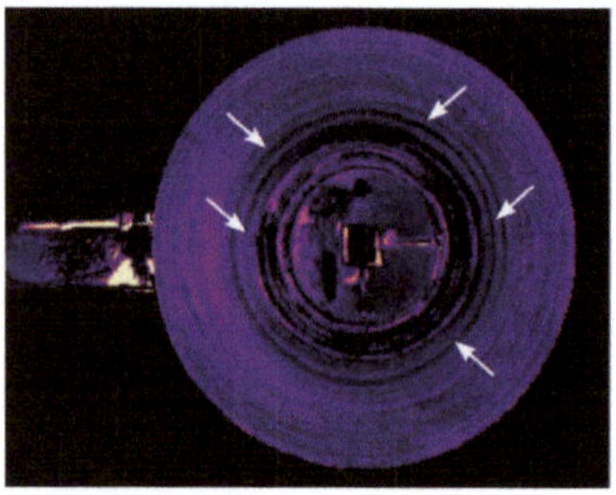

(b) $t = 185.0$ s

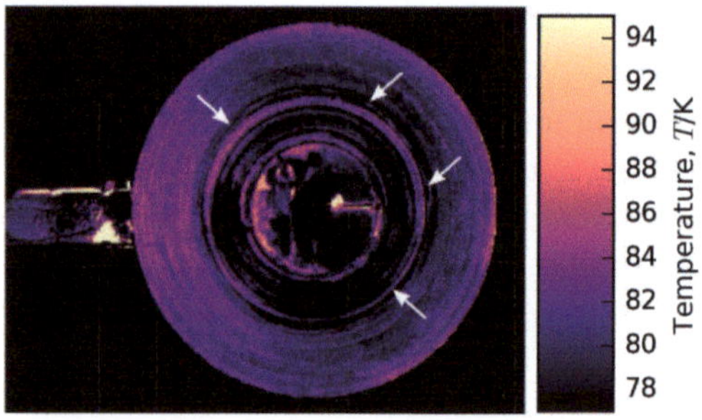

(c) $t = 297.80$ s

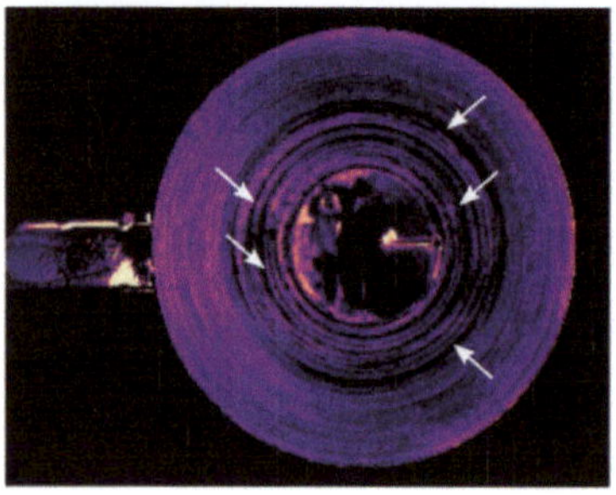

(d) $t = 298.80$ s

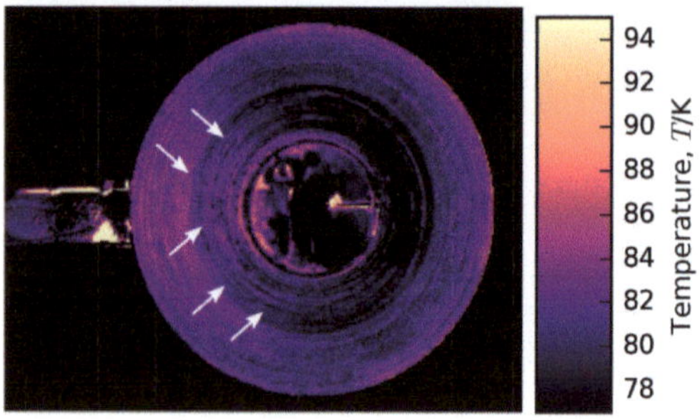

(e) $t = 300.0$ s

Figure 5.26: Electrical and thermal measurement results of the 72 A overcurrent measurement. Dashed lines indicate positions of thermal images. (a) shows the time development of the applied current ramp, coil voltage and magnetic field. (b)-(e) show thermal images, where white arrows indicate specific features.

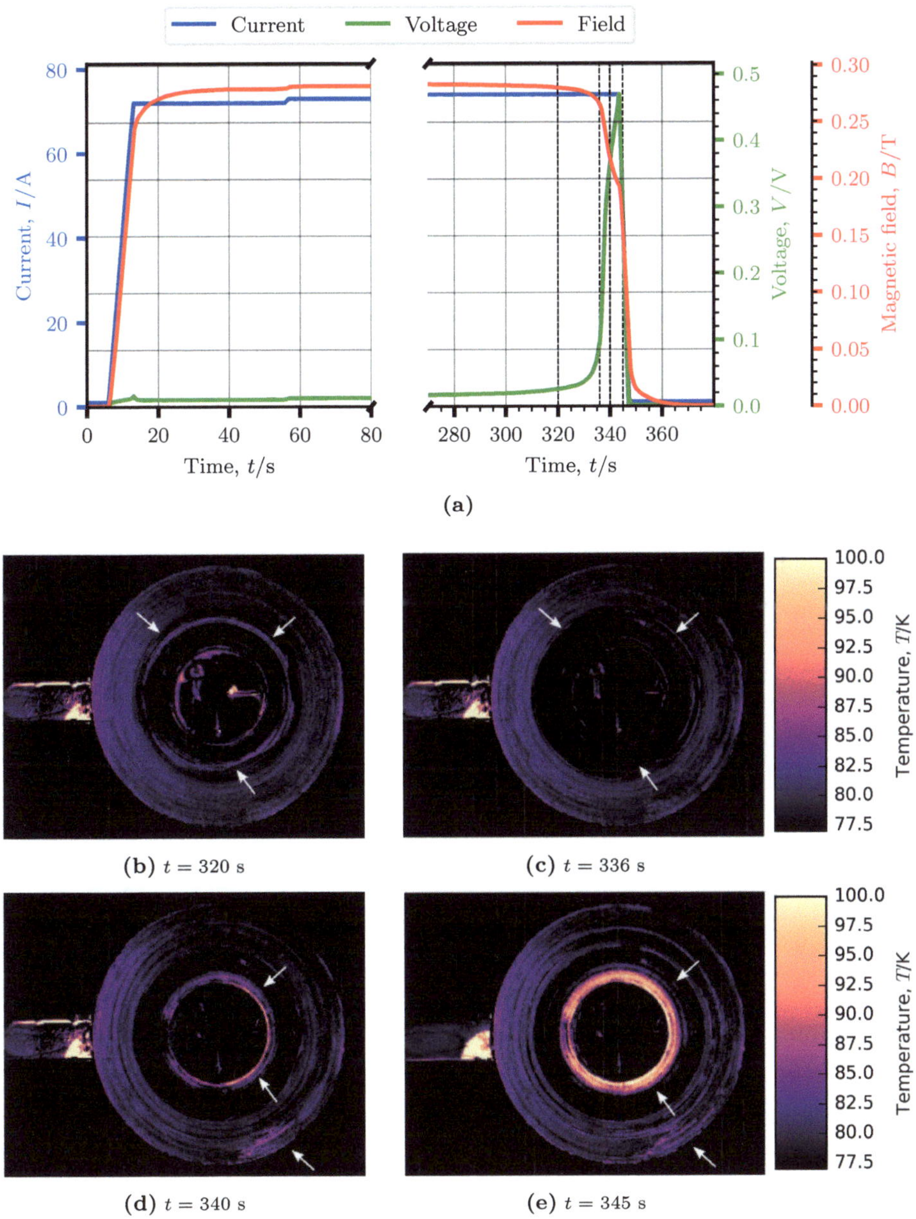

Figure 5.27: Electrical and thermal measurement results of the 74 A overcurrent measurement. Dashed lines indicate positions of thermal images. (a) shows the time development of the applied current ramp, coil voltage and magnetic field. (b)-(e) show thermal images, where white arrows indicate specific features.

5.4.7 AC current ripples

Most dc power supplies have a certain current ripple on the output either coming from the grid frequency or due to the rectification of the current. In rotating machines the changing magnetic field induces ac current in conductors that have in principle a similar effect. Such ripples often appear at higher multiples of the power line frequency and are commonly 1 % to 3 % of the nominal current. For these reasons the effect of ac current ripples was also investigated on the test coil using various frequencies and ripple amplitudes, generated according to eq. 5.22.

$$i(t) = I + I_{\text{ripple}} \cdot \sin(2\pi f t) \tag{5.22}$$

where I was a steady dc current between 30 A and 50 A, I_{ripple} was between 1 A and 10 A and the frequencies, f, ranged from 0.1 Hz to 400 Hz. The measurements were recorded both via fluorescent thermal imaging as well as electrically, making sure that the data acquisition speed was fast enough to capture high frequency signals with satisfactory resolution. By monitoring the change of magnetic field at the centre of the coil simultaneously to currents, the actual amount of current bypassing the coil – and as such flowing through the radial path – was determined using the coil field coefficient. Table 5.2 shows the measured voltage and magnetic field peak-to-peak fluctuations with a 1 A and 10 A ac current superimposed on top of a 30 A dc current in the coil with frequencies from 0.5 Hz to 400 Hz. Figure 5.28 then shows the fluctuations of the magnetic field with only the 10 A ac signal.

The variations in field are too weak in the case of the 1 A ac signal, however they are clearly visible with the 10 A ac current waveform. With the coil constant of 3.98 mT A^{-1}, a 20 A peak-to-peak signal – if completely following the spiral path inside the coil – would generate magnetic field fluctuations of

$$k \cdot i_{\text{max}} = 0.0796\,\text{T} \tag{5.23}$$

in the measurement. Any lower change in field then indicates that the current is proportionally bypassing the spiral path and flows through the turn-to-turn contact resistances to some extent. The results in Table 5.2 indicate that this is the case in most higher frequencies. Even at 1 Hz the field change is merely 32 mT, representing about 53 % of the change, in case if the current were to completely follow the spiral path. However, this is not unexpected, since even this frequency already represents an effective ramping rate of 40 A s^{-1}.

Further experiments using the same principle of dc offset current, as high as 50 A ($\sim$80% I_c), with modulated ac signals up to 100 Hz in frequency and 1 A, 2 A and 10 A

Table 5.2: Peak-to-peak variations in central magnetic field and induced voltage with a 1 A and 10 A ac signal modulated on top of a 30 A dc current, for frequencies from 0.5 Hz to 400 Hz. † Shows the percentage of the ac current part bypassing the spiral path completely as calculated from the coil field coefficient.

Frequency	1 A ac current		10 A ac current		
(Hz)	$\Delta B_{\mathbf{max}}$ (mT)	$V_{\mathbf{max}}$ (mV)	$\Delta B_{\mathbf{max}}$ (mT)	$V_{\mathbf{max}}$ (mV)	† Bypass (%)
0.5	-	-	41.66	14.7	47.66
1	3.09	2.2	32.34	18.9	59.37
5	1.39	2.6	12.80	22.6	83.92
10	9.22	2.5	9.32	23.9	88.29
25	1.12	3.1	6.56	25.1	91.76
50	0.91	3.1	4.88	25.6	93.87
100	1.18	3.1	3.46	26.7	95.65
200	0.78	3.1	2.35	27.4	97.05
400	0.68	3.0	1.30	25.7	98.37

in amplitude are shown in Figure 5.29. The coil was ramped with 5 A s^{-1} to the target current, where the ac current was applied at 1 A with frequencies from 0.1 Hz to 100 Hz. After this the ac signal amplitude was increased and the signals repeated. At the end the coil was ramped down using the same ramping rate as during charging.

It was found that the 1 A ac current had no effect on the coil, however at 2 A during the 25 Hz ac current, a visible hot zone was detected in the coil as shown in Figure 5.29b and Figure 5.29c. The heating around the coil's centre (indicated by the white arrows) was already present after ramping the coil. A hot zone on the upper right side of the coil appeared during the 25 Hz signal. This was shortly followed by 50 Hz and 100 Hz ac currents and as such the warmer region of the coil did not recover and a temperature offset was visible until the end of the signals. After this, the warmer region started to recover. No heating was observed during ac signals with 10 A in amplitude for frequencies between 0.1 Hz to 5 Hz and the coil was slowly recovering until the ramp down as shown in Figure 5.29d.

At the exact time when the coil discharging began at the end of the experiment, a brief yet distinctly visible heating was detected once again in the same zone of the coil as before, shown in Figure 5.29e.

Due to the limited thermal imaging recording length, the 10 A 25 Hz, 50 Hz and 100 Hz measurements were done in a subsequent measurement. While this measurement was

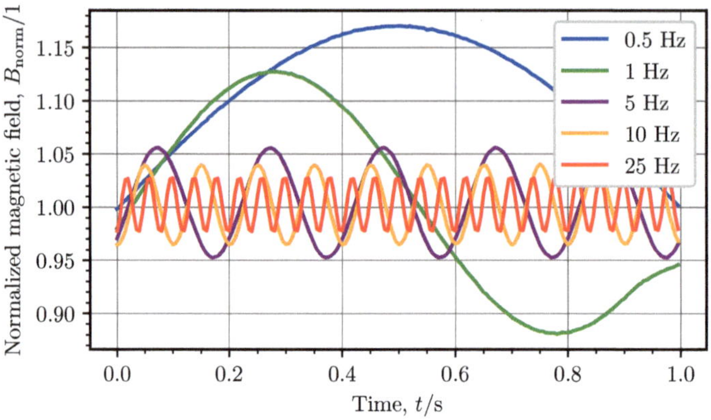

(a)

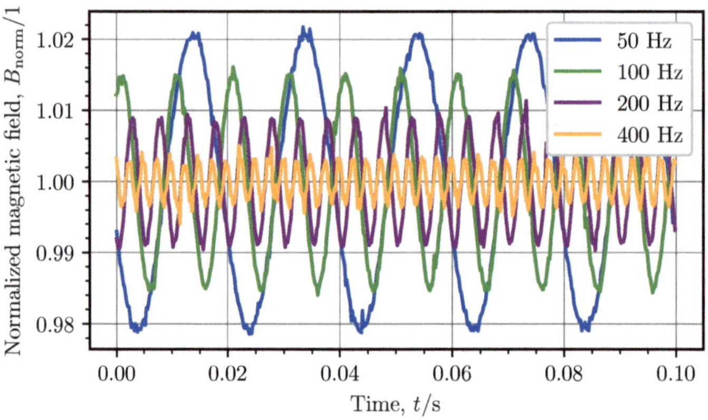

(b)

Figure 5.28: Normalized magnetic field during an alternating current with an amplitude of 10 A applied on top of a steady 30 A dc current. (a) shows frequencies between 0.5 Hz to 25 Hz and (b) frequencies between 50 Hz to 400 Hz. In (b) the power supply experienced initially a small drop in the dc offset current, which was manually corrected for in this plot.

merely a few minutes after the previous one, no warmer zones were observed with only minor heating during the experiment. A sudden heating at the beginning of the ramp down was also not observed in this measurement.

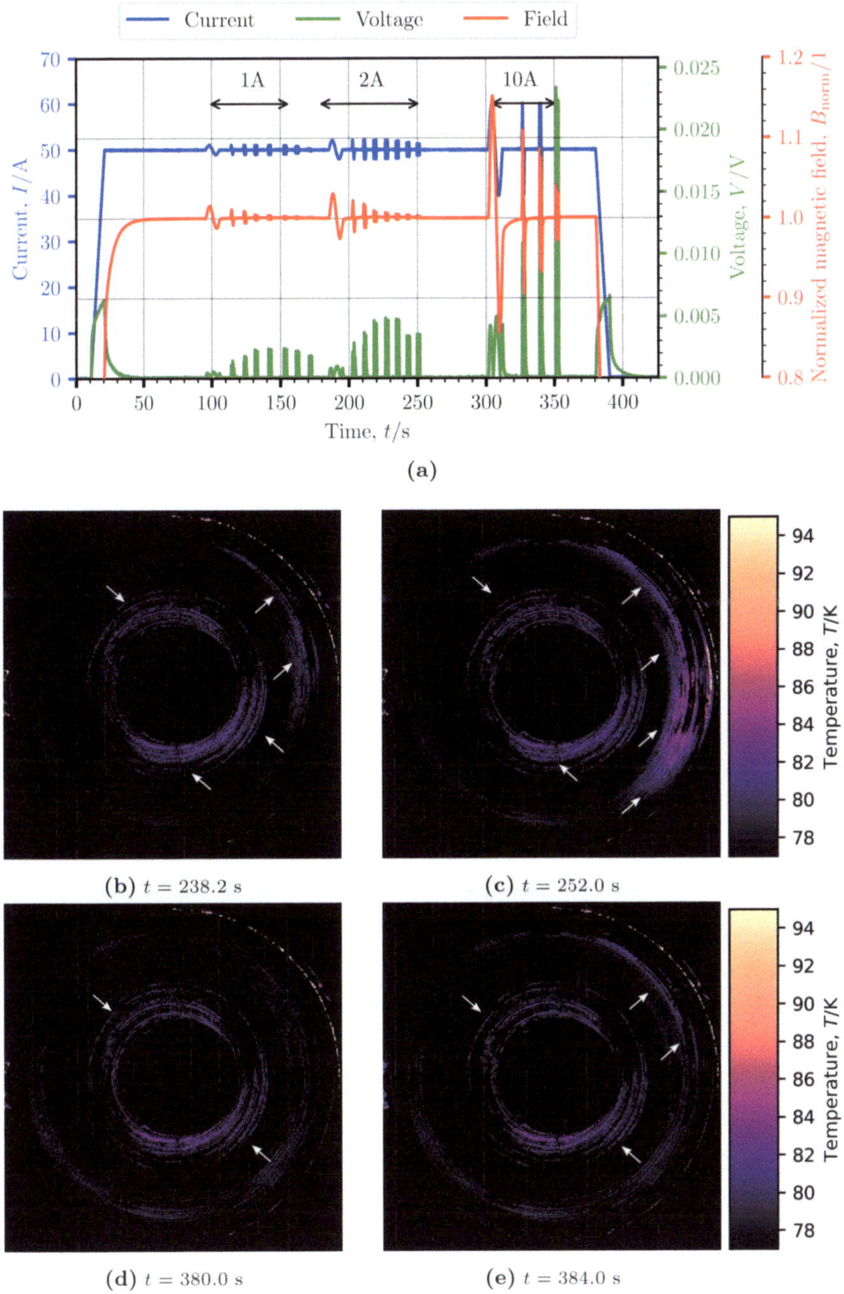

Figure 5.29: Electrical and thermal measurement results of 1 A, 2 A and 10 A ac current from 0.1 Hz to 100 Hz on top of a steady 50 A dc current. (a) shows the electrical measurements with a normalized magnetic field. Arrows indicate the amplitude of the applied ac current. (b)-(c) show initial heating during the 2 A ac signal between frequencies 25 Hz to 100 Hz. (d)-(e) show the coil partially recovered and sudden, local heating at the moment of ramping down.

5.5 Insulated coil measurement results

5.5.1 Critical current

The insulated coil was initially also measured in liquid nitrogen to obtain the critical current of the coil. The results are shown in Figure 5.30, with the critical current determined as 65 A and 72 A using the 0.1 µV cm^{-1} and 1 µV cm^{-1} criterion, respectively. Due to the added thickness of the Kapton® insulation between the turns, the magnetic field generated by the coil is reduced and hence the critical current is marginally higher, compared to the NI coil. After measuring the coil's critical current in liquid nitrogen and confirming the characteristic properties, further measurements were carried out in the conduction-cooled experimental assembly.

5.5.2 Thermal and electrical effects of various ramping speeds

The coil was ramped with 0.1 A s^{-1}, 0.5 A s^{-1}, 1 A s^{-1}, 2 A s^{-1}, 5 A s^{-1} and 10 A s^{-1} ramping rates until 40 A, 50 A and 60 A while both thermal images as well as the magnetic field and coil voltage development were recorded. Unfortunately, despite the more accurate magnet ramping current source, due to the inductance of the coil and the minor current fluctuations, the noise in the fast DAQ system was still too high to do reliable measurements. Hence, instead of the DAQ signal, the nanovoltmeter was used at a higher speed with the readout range set to 1 V for sufficient accuracy.

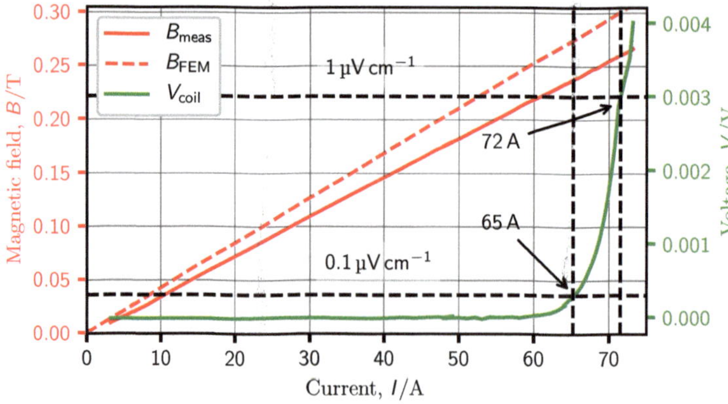

Figure 5.30: Initial I_c measurement of the insulated coil showing the 0.1 µV cm^{-1} and 1 µV cm^{-1} electric field criterion. The limiting critical current of the coil is then 65 A and 72 A, respectively. The plot shows data points extracted from the measurement and not the actual ramping plot.

The inductance of the coil was determined using eq. 2.11 and averaging over several ramping measurements was determined as 1.44 mH.

Figure 5.31 shows the magnetic field development of the coil with varying current ramping rates. Compared to the NI coil (refer to Figure 5.17) it is visible that this coil has no charging delay, the magnetic field is directly proportional to the applied current.

Due to the zero resistance path of the superconducting windings and the lack of turn-to-turn resistive bypass paths, no distinct heating was observed during the ramping measurements. The used power supply did not allow the generation of ac waveforms and the use of extended overcurrent was also avoided due to accidental burnout. Instead, a different approach was used to try and determine the stability of this coil.

5.5.3 Heat pulses

As the coil voltage rose relatively quickly around current levels of 70 A, going considerably above the critical current was avoided in ramping measurements. Instead, to detect any anomalies in the coil and observe its behaviour in transient states, a heater was added on the surface and short heat pulses were applied. This was previously introduced in Section 5.3.5 and shown in Figure 5.12.

The heat pulses were carried out using varying heating power, duration and at different current levels. The three current levels used in the coil were 40 A, 50 A and 60 A and the heater was pulsed with approximately 1.25 W and 3.2 W for durations between 1 s to 7 s. In the measurements the coil was ramped using a rate of 5 A s^{-1}, where upon

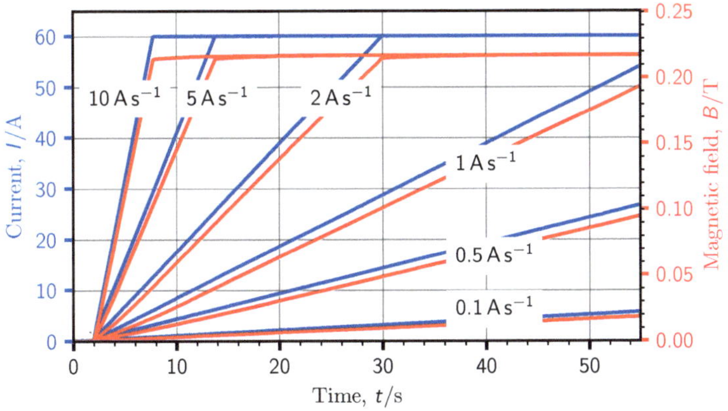

Figure 5.31: Time development of current and magnetic field in the insulated coil when ramped with a rate between 0.1 A s^{-1} to 10 A s^{-1} until 60 A.

reaching the target current a 60 s stabilization period was allowed. Thereafter, the heater was pulsed with the selected power for durations incremented by 1 s at a time. After each heater pulse the coil was ramped down using the same 5 $A\,s^{-1}$ ramping rate.

Heating before detectable voltage

In the measurements up to 60 A coil current and heat pulses of up to 2 s at a power of 3.2 W no voltage was detected between the coil's terminals. Figure 5.32 shows the results of this experiment where the heater was pulsed for a duration of 2 s with a nominal current of 800 mA. As visible in Figure 5.32a there was no detectable voltage signal in the coil. However, looking at Figure 5.32b and Figure 5.32c, a minor amount of heat transfer is already visible from the heater into the coil. While this is indeed barely visible in the static image, in the time development of the post-processed measurement the signal can be identified. This indicates that while heat disturbances were visible, these do not necessarily cause magnetic field fluctuations or any voltage response, provided that the local critical current density of the coil is not exceeded by the transport current.

Normal transition

In the heater pulse measurements the coil remained stable until heater current pulses of 800 mA (3.2 W) for a duration of 6 s, with a total heat dissipation of 19.2 J in the resistor. While a clear voltage drop was measured, the coil recovered after each heat pulse ended. A normal transition was first observed when the pulse duration was increased to 7 s, providing about 3.2 W of heating power and an overall energy of 21.4 J. This measurement is shown in Figure 5.33a. The heater was started just before 81 s, which is visible in Figure 5.33b as the heating of the resistor changes the reflected light and hence causes a false thermal reading, nevertheless, useful for timing the heater. Approximately 2.5 s after the heater was started, a voltage signal was also detected. However, after turning the heater off, the coil did not recover and a thermal and electrical runaway began. This is visible in Figure 5.33c, where the heater was already off, yet the temperature kept increasing. Figure 5.33d then shows that the heating is concentrated predominantly to the turns touched by the heater and that propagation to neighbouring turns is negligible. In this image one can see a hot ring reaching temperatures of 150 K. At this point the sharp transitional voltage caused an overflow in the nanovoltmeter and simultaneously caused the power supply to trigger quench protection and drop the current to zero. As such, the coil was protected and the heat started to diffuse further along the turn as well as into neighbouring windings as shown in Figure 5.33e.

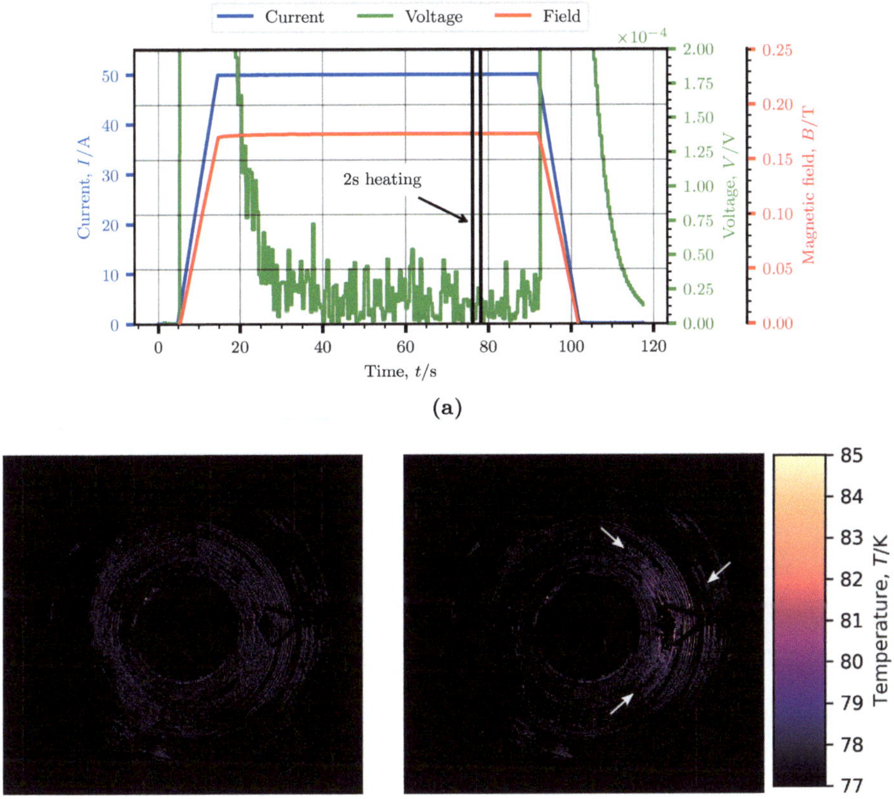

(a)

(b) $t = 75$ s (c) $t = 79.5$ s

Figure 5.32: Electrical and thermal measurement results of a 2.5 s long heater pulse in the insulated coil. (a) shows the time development of the applied current ramp, coil voltage, magnetic field as well as the heater pulse. (b) and (c) show thermal images, where white arrows indicate specific features.

Unlike in the case of the NI coil, where the normal transition was slow and accompanied by a detectable reduction of the centre magnetic field, in this insulated coil the field remained unchanged even with increasing voltage. A lack of electrical bypass paths led to large voltages developed and quasi-local heating over the involved windings, which – if undetected – may have destructive consequences.

After the quench, the coil's critical current was re-measured and was found to be unchanged.

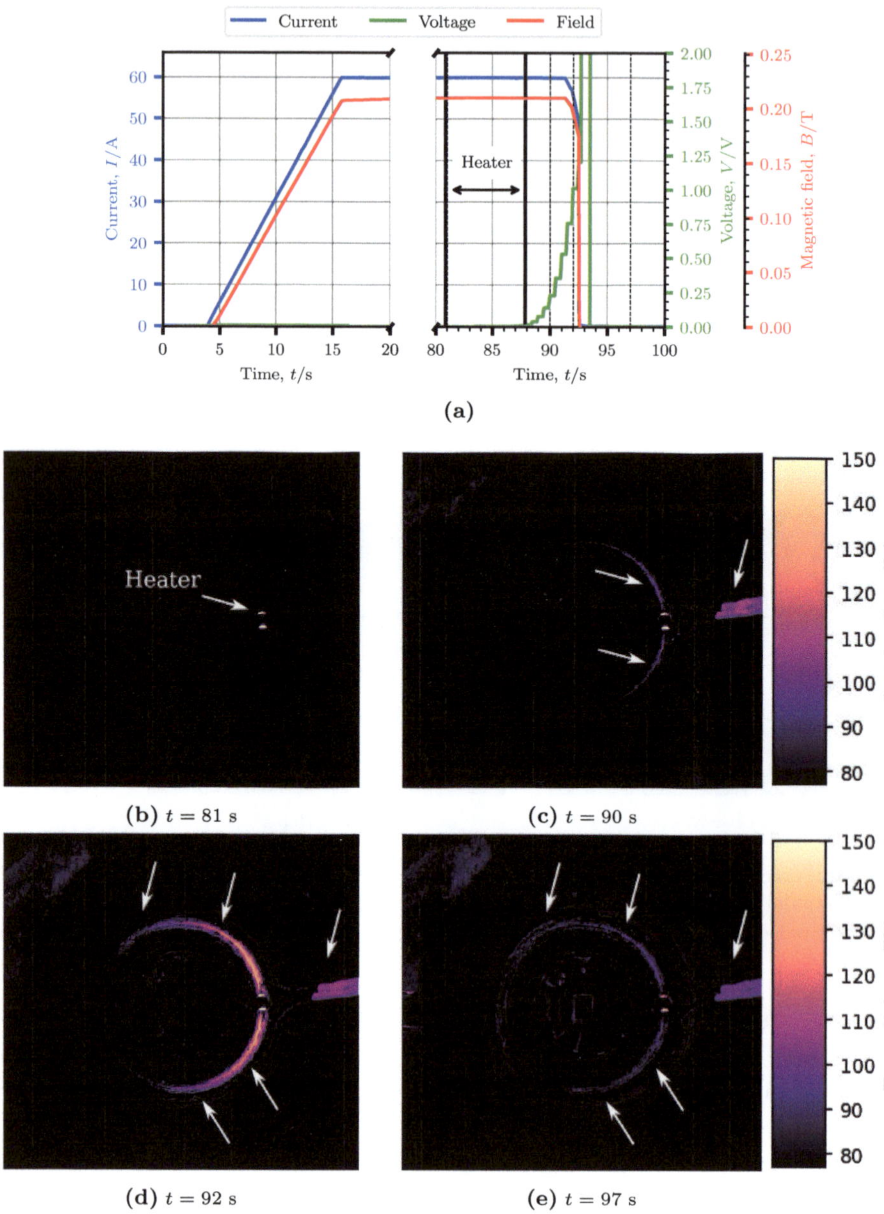

Figure 5.33: Electrical and thermal measurement results of the 7 s long heater pulse in the insulated coil. Dashed lines indicate positions of thermal images. (a) shows the time development of the applied current ramp, coil voltage, magnetic field as well as the heater pulse. (b)-(e) show thermal images, where white arrows highlight specific features. Colour changes of the heater are optical artefacts and should be ignored.

5.6 Summary

Two pancake coils were wound from HTS coated conductors into an approximately 3 cm inner and 10 cm outer diameter coil. The first coil was wound without turn-to-turn insulation whereas the second coil used Kapton® as an insulating layer between the turns. The surface of both coils was coated with the fluorescent EuTFC-based coating with the aim to observe local heating in the conduction-cooled assembly during various operating phases.

5.6.1 Non-insulated coil

Electrical measurements on the NI coil have shown than when ramping with current rates between $0.1\,\mathrm{A\,s^{-1}}$ to $100\,\mathrm{A\,s^{-1}}$ to 40 A, 50 A and 55 A (60 %, 78 % and 85 % of the coil's I_c respectively) the coil charging delay is a function of the ramping rate, where a faster ramping rate increases the charging delay. This delay, however, became increasingly saturated above $5\,\mathrm{A\,s^{-1}}$. At this charging rate the coil needed approximately 16 s to reach 99.3 % of the target magnetic field from the moment where the ramping has ended. From sudden discharge measurements it was also determined that the discharge time constant of the coil is 610 ms and it is not dependent on the current amplitude from where the coil was discharged.

Fluorescent thermal imaging measurements have not revealed a consistent increase in heating at increased ramping rates. While heating was present in some parts of the coil during some of the ramps, due to the lack of repeatability these may be called anomalies related to the cooling performance inside the experimental assembly or some other undetermined factor. Similar results were observed in the case of the sudden discharge measurements, where no reasonable heating was detected. Due to the size and stored insignificant amount of energy in the coil, no heating was present in a sudden discharge. It is believed that larger coils with considerably higher inductance would behave differently in many scenarios when compared to the small test coil prepared in this work.

Overcurrent measurements of the coil revealed an anomalous behaviour, where a continuously increasing coil voltage suddenly recovered. In the measurement the magnetic field started dropping, accompanied by an increase in coil voltage, both signalling an increasing current transfer into the radial path. However, instead of a catastrophic thermal runaway, the coil current redistributed and both voltage and magnetic field levels returned to normal. The effects were visible on the thermal imaging, which revealed rapid current redistribution (and heating redistribution) at the moment of sudden recovery.

In yet another overcurrent measurement the coil was ramped to 74 A (110 % of the critical current) and kept there until the beginning of a thermal runaway. The thermal imaging measurements clearly showed that the origin of the normal zone was at the centre of the coil, where a resistive joint exists between coil former and superconductor. Simultaneously, this is the point of highest magnetic field. Over the duration of several seconds, only the few innermost turns heated up to approximately 100 K. The coil was manually protected by ramping it down and measuring its I_c after the quench revealed no degradation. The measurement has shown that the thermal runaway was virtually constrained to the location where it has started with little to no propagation in the radial direction.

Applied ac current to the coil, modulated on top of a steady dc current, mostly bypassed the spiral path and hence caused only minor fluctuations in the magnetic field. Heating due to the ac current was also not visible at ac amplitudes of 1 A and 10 A on a 30 A dc current due to the low characteristic coil resistance. A 10 A current ripple (which is already significant 15 % of the coil's I_c) would cause merely 0.2 W of heating when approximated as I^2R losses with a characteristic coil resistance of 2.23 mΩ, also assuming a complete and uniform bypass of the spiral path. Sudden heating as a result of a 2 A 25 Hz to 100 Hz ac current on top of a 50 A dc current became visible in one of the measurements. In the same measurement, the coil has shown local heating in the same position when ramping the coil down. Interestingly, however, this behaviour could not be reproduced in a subsequent measurement with the same frequencies using a higher, 10 A ac signal. The local heat generation suggests that ac current bypass may not happen completely uniformly in a coil, instead it could be concentrated to parts of the coil with slightly different turn-to-turn resistance characteristics.

5.6.2 Insulated coil

In the prepared insulated coil, no distinct or characteristic heating was observed when ramped from $0.1\,\mathrm{A\,s^{-1}}$ to $10\,\mathrm{A\,s^{-1}}$ until different current levels. To avoid accidentally burning out the coil in an overcurrent measurement, instead a resistive heater was mounted on the coil surface with the aim to observe the stability and the normal transition behaviour. The coil withstood a continuous heat input of 3.2 W for up to 6 s, where both the electrical and thermal signs of instability were observable. Nevertheless the coil recovered on its own after the heater was turned off. When the heater pulse duration was increased to 7 s, a thermal runaway began. Starting from the heater's position a hot zone developed, which spread predominantly along the turns in contact with the heater in a semi-circle shape. Neighbouring turns were mostly unaffected by the thermal runaway. The coil was automatically protected by the current supply's built-in quench protection mechanism. Measuring the coil's critical current after the thermal runaway did not reveal any damage in this case either.

6 Conclusion and outlook

In this work a 2D temperature mapping method was developed for quench analysis in HTS tapes and applications. The application of high-speed fluorescent thermal imaging was demonstrated at 77 K on several HTS tapes as well as two HTS pancake coils, one of which was wound as a non-insulated coil.

Various HTS tapes were measured with pulsed currents to observe quench development and propagation in liquid nitrogen caused by a magnetic defect. The conductors varied in manufacturer, stabilization as well as architecture, including the novel current flow diverter concept. By defining the normal zone from a temperature perspective as areas above the critical temperature of HTS, a normal zone propagation velocity could also be calculated from the thermal imaging method. Cross-validating the measurement results with data provided on identical tapes by Polytechnique Montréal, Canada revealed that at higher operating currents the NZPVs calculated from electrical and thermal readings match more accurately than at lower currents. In the latter case the thermal imaging consistently reported slower NZPV, with up to one order of magnitude difference. This mismatch between the NZPVs was ascribed to the quadratic dependence of heating power on the applied current, $P = I^2 \cdot R_{sc}$. While a fast electric field development was possible in HTS tapes with low critical currents, the low heating power in these tapes caused the thermal zone propagation to lag the electrical signal. On the other hand, in tapes with a higher critical current (and hence higher applied currents), the considerably higher heating power caused the propagation of the thermal zone to match closely that of the electric field.

The appearance of multiple quench zones was visible in several tapes at current pulses slightly above the critical current. The appearance of these additional quench zones followed the initial quench almost instantaneously and the behaviour is ascribed to the presence of natural defects. The existence of a quench in this manner may not be immediately obvious from electrical measurements and indicates that the concept of quench and normal zone propagation in HTS is not necessarily as straightforward as a single propagating zone at a given speed. The presence of additional quench zones have a big influence on the heating distribution in the tapes.

The test coils were measured using various current ramping rates and current levels. Due to a low coil resistance and critical current, even a fast ramping rate of 100 $A\,s^{-1}$

did not cause any visible heating. However, this is most likely the consequence of a small demonstrator coil. In a larger coil, where the ramping would take considerably longer, resistive joints as well as currents bypassing through the turn-to-turn resistances would be expected to cause higher losses and possible heating. Investigating ac current ripples on top of a steady coil current in the coil yielded the same results, where the low currents involved made it difficult to observe any heating related to the ac current.

The NI coil was also investigated in overcurrent scenarios, where the coil was ramped until 72 A and 74 A (corresponding to 108 % and 111 % of the coil's critical current) and held steady to observe a thermal and electrical runaway. In the 72 A measurement a voltage shift was observed across the coil's terminals, which was accompanied by a steady drop of the central magnetic field over the duration of several seconds. This was a clear indication of an increasing fraction of the current bypassing the spiral path and flowing across the turn-to-turn resistances. At one point, however – without any intervention – the coil voltage fell back to previous levels and the magnetic field also increased in a step-like manner. On the thermal images a clear redistribution of the heating load was visible at the time of the sudden change. The exact reason for this behaviour is currently not clear and requires further investigation.

In a subsequent measurement the NI coil was ramped to the previously validated current level, however at this time the coil remained stable. An instability became only apparent when slowly ramped to 74 A, where the coil was held for a longer period. During this measurement the coil voltage increased at a slow, yet steady rate and eventually a runaway was recorded where the coil was quickly ramped down. On the thermal image clear heating was visible around the centre windings of the coil, which – over the course of a few seconds – did not spread in the axial direction in the body of the coil, but remained constrained to the innermost few windings. Temperatures up to 100 K were measured at this position. This behaviour was expected in an NI coil without any artificial defect, since the centre was the position of highest magnetic field coinciding with a resistive joint in the design.

The results indicate that while NI coils are highly stable, their behaviour is not completely understood yet. While no natural weak spots were detected during operation, in the case of an overcurrent the heating did occur locally and was not distributed rapidly.

Similar to the NI coil, the insulated coil was also thermally imaged in operation, where no reproducible heating was visible in the coil during various ramping speeds. Instead, to test the stability of the coil and the capabilities of the thermal imaging method, a resistive heater was mounted on the surface of the coil (retrofitted) and current pulses of varying length and amplitude were applied to the heater. With increasing heater pulses the coil was destabilized using a heater power of 3.2 W for 7 s at a dc coil current of 60 A. Here a local hot spot development was observed after the heater pulse at the position of the heat input and a temperature of 150 K was measured before the coil was protected. The heating was concentrated to the initiation of quench and did not spread in the transverse direction. A temperature rise was visible along the turns in a half-circle shape, constrained to the windings initially disturbed by the heater.

While the wound demonstrators coils have shown no heating due to local defects or reasonable heating during various ramping measurements, this may be due to the involved low coil resistance and current levels. Larger coils with significantly higher inductance would incur a higher heat load during ramping and therefore are expected to behave differently.

The introduced method has the potential for measuring temperatures in superconducting applications. Beyond the work presented in this thesis, several improvements are possible that would make the technique more applicable and useful for specific scenarios. While the current coating technique was suitable for flat tapes, coating uneven surfaces – even that of the coil – proved challenging. Researching another method could make the coating process simpler and more reliable, however, it would most likely require a novel solvent for the Europium fluorescent dye, which would require further investigation regarding performance at cryogenic temperatures and fluorescent properties. In this work, due to the required fast, transient effects to be recorded, a high-speed camera was used. The working principles of this camera prevented the method to be used in real-time. Simultaneously, due to the different target market of the used camera, it has incurred in significant noise at low light intensities. By the choice of the camera (for example higher digital resolution), the method could be made more appropriate for specific measurements where accuracy is of higher importance than acquisition speed. In this case, the method could also be implemented in true real-time using simultaneous post-processing.

A Appendix

Figure A.1 shows the 3D-printed frame for the optical filter. Since such narrow bandpass filters are commonly made for optics applications with given standard sizes and mounting options, finding these in larger diameter for a camera lens front-mount proved difficult. Using a custom-made frame allowed using a standard optical filter in between the camera lens and the camera's CMOS sensor and enabled blocking a wide spectrum of unwanted light.

UV lights for microscopy measurements often use light beams that illuminate a small surface of only a few mm^2. Commercial UV light fixtures used for curing, on the other hand, while precise in emission spectrum are often extremely expensive and bulky. As a compromise, an array of LEDs was used as shown in Figure A.2 built from an aluminium heat sink and several UV LEDs already mounted on a starboard PCB.

The heat transfer between a submerged object and boiling liquid nitrogen at 77 K is highly non-linear as shown in Figure A.3. The optimal heat transfer is between

Figure A.1: The 3D-printed frame allowing to insert the optical bandpass filter in between the camera and the lens.

Figure A.2: Array of LEDs on a starboard PCB used as excitation light in the experiments. The LEDs were connected in series and driven with their nominal current of 700 mA.

10 K to 20 K temperature difference. In practical applications and during a quench of superconductors, exceeding this limit severely limits the cooling performance due to the formation of a thin film boiling region that acts as an insulation.

Figure A.4 shows the dependence of the n-value on the applied magnetic field and angle [122]. This data was used in the commercial FEM software COMSOL Multiphysics to calculate the coils' critical current.

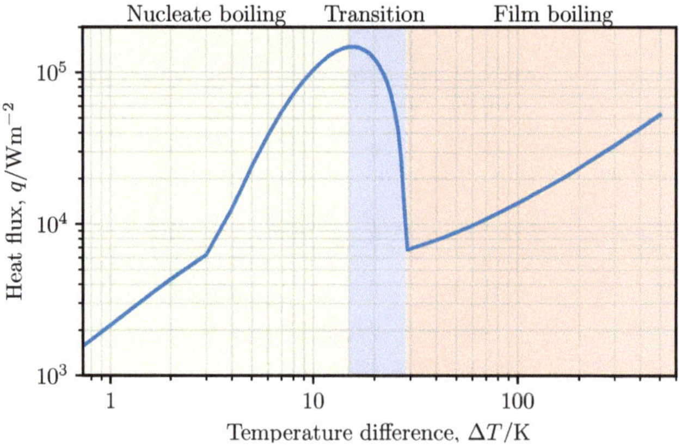

Figure A.3: Typical heat flux between liquid nitrogen at a temperature of 77.3 K and a submerged object [106].

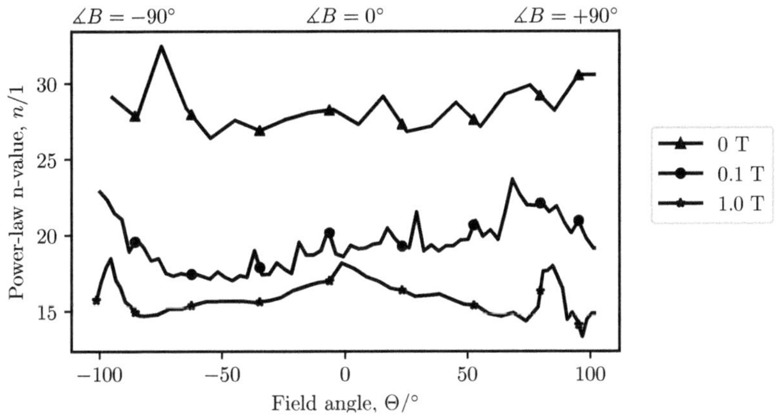

Figure A.4: Dependence of the power-law n-value on the magnetic field angle and amplitude, used in simulations [122].

List of Figures

List of Tables

List of Symbols

Overview of HTS tapes and their stability

κ_{cd}	Thermal conductivity,	$W\,m^{-1}\,K^{-1}$
μ	Permeability of materials,	$H\,m^{-1}$
ω	Angular frequency,	$rad\,s^{-1}$
ρ_{cd}	Electrical resistivity of a conductor,	$\Omega\,m$
ρ_{ct}	Turn-to-turn contact resistivity,	$\Omega\,cm^2$
ρ_{sc}	Electrical resistivity of a superconductor,	$\Omega\,m$
τ	Universal time constant,	s
τ_{down}	Coil discharge time constant,	s
τ_{up}	Coil charging time constant,	s
Θ	Angle of incidence of the magnetic filed,	$^\circ$
A	Cross-section area of a conductor,	m^2
A_t	Cross-section area of an HTS tape,	m^2
B, $\mathbf{B}$	Magnetic flux density,	T
B_{c1}	Lower critical magnetic field,	T
B_{c2}	Upper critical magnetic field,	T
B_c	Critical magnetic field,	T
B_{max}	Maximum magnetic field,	T

e	Base of the natural logarithm,	2.718 28
E, $\mathbf{E}$	Electric field,	$\mathrm{V\,m^{-1}}$
E_{c}	Electric field criterion,	$0.1\,\mathrm{\mu V\,m^{-1}}$ and $1\,\mathrm{\mu V\,m^{-1}}$
E_{sc}	Electric field in the superconductor,	$\mathrm{V\,m^{-1}}$
h	Heat transfer coefficient,	$\mathrm{W\,m^{-2}\,K^{-1}}$
H, $\mathbf{H}$	Magnetic field strength,	$\mathrm{A\,m^{-1}}$
H_{c}	Critical magnetic field strength,	$\mathrm{A\,m^{-1}}$
I	Direct current,	A
$i(t)$	Alternating current,	A
$I_{\mathrm{c,op}}$	Critical current at operating temperature,	A
I_{c}	Critical current,	A
I_{r}	Current in the radial path,	A
I_{source}	Supply current,	A
I_{s}	Current in the spiral path,	A
J, $\mathbf{J}$	Electrical current density,	$\mathrm{A\,m^{-2}}$
J_{c}	Critical current density,	$\mathrm{A\,m^{-2}}$
L	Inductance,	H
l_{mpz}	Minimum propagating zone length,	m
n	Power law exponent,	-
N_{t}	Number of turns,	-
Q_{MQE}	Minimum quench energy,	J or $\mathrm{J\,cm^{-3}}$
R	Electrical resistance,	Ω
R_{ct}	Turn-to-turn contact resistance,	Ω

R_{c}	Characteristic coil resistance,	Ω
R_{i}	Resistance across the ith turn of the coil,	Ω
r_{i}	Radius of the ith turn of the coil,	m
R_{sc}	Resistance of a superconductor,	Ω
R_{shunt}	Shunt resistance,	Ω
S_{t}	Perimeter of an HTS tape,	m
T	Temperature,	K
T_{cs}	Current sharing temperature,	K
T_{c}	Critical temperature,	K
t_{delay}	Coil charging delay,	s
T_{op}	Operating temperature,	K
V	Electric potential,	V
V_{coil}	Electric potential across the coil terminals,	V
w_{d}	Width of a superconducting tape,	m

A high-speed thermal imaging method for cryogenic temperatures

α	Absorption,	-
ϵ	Emissivity,	-
η	Optical collection efficiency,	-
λ	Wavelength,	m
λ_0	Central wavelength,	m
λ_θ	Wavelength at a given angle,	m
σ	Stefan-Boltzmann constant,	$5.670\,373 \times 10^{-8}\,\mathrm{W\,m^{-2}\,K^{-4}}$
τ	Transmissivity,	-

$\varepsilon_{\mathrm{rel}}$	Relative error,	-
A	Surface area,	m^2
b	Wien's displacement constant,	$2.897\,772\,9 \times 10^{-3}\,\mathrm{K\,m}$
I	Excitation light intensity,	-
n	Refractive index of a given medium,	-
n_*	Refractive index of the filter,	-
P	Radiated power,	W
Q	Quantum efficiency,	-
r	Reflectivity,	-
S	Light intensity,	-
S_{C}	Light intensity at some low temperature (unbiased),	-
S_{H}	Light intensity at some higher temperature (under bias),	-
S_{R}	Light intensity ratio,	-
$\overline{S}$	Mean light intensity,	-

Quench propagation and analysis in HTS tapes

I	Direct current,	A
P	Electrical power,	W
Q	Energy,	J
t	Time,	s
t_{p}	Current pulse duration,	s
x	Distance,	m

Fluorescent thermal imaging of HTS pancake coils

Ω	Boundary of a superconducting domain,	-

σ	Electrical conductivity,	$\mathrm{S\,m^{-1}}$
σ_{max}	Maximum electrical conductivity in the FEM model,	$1 \times 10^{16}\,\mathrm{S\,m^{-1}}$
σ_{norm}	Normal state electrical conductivity in the FEM model,	$1 \times 10^{6}\,\mathrm{S\,m^{-1}}$
σ_{sc}	Electrical conductivity of the superconductor,	$\mathrm{S\,m^{-1}}$
Θ	Magnetic field angle,	$^{\circ}$
$\mathbf{A}$	Magnetic vector potential,	$\mathrm{V\,s\,m^{-1}}$
$\mathbf{n}$	Normal vector,	-
e_{coil}	Energy stored in the coil,	J
I_{app}	Applied current,	A
$I_{\mathrm{c,coil}}$	Critical current of the coil,	A
$I_{\mathrm{c,mean}}$	Mean critical current of the coil,	A
$I_{\mathrm{c,min}}$	Minimum critical current of the coil,	A
$I_{\mathrm{c,tape}}$	Critical current of the tape,	A
I_{max}	Coil current at peak magnetic field,	A
i_{max}	Peak-to-peak alternating current amplitude,	A
I_{ripple}	Amplitude of the applied dc ripple current,	A
k	Coil magnetic field constant,	$\mathrm{T\,A^{-1}}$
m_{coil}	Coil mass,	kg
n	Power law exponent,	-
n_{t}	Selected single turn of a coil,	-
Q_{coil}	Energy absorbed by the coil,	J
t_{tape}	Tape thickness,	m

Bibliography

[1] Power transformers Built for reliability and efficiency. Technical report, ABB, 2011. (Visited on 21/02/2019) (cited on page 1).

[2] U.S. Energy Information Agency. 2019. URL: https://www.eia.gov/tools/faqs/faq.php?id=105&t=3 (visited on 21/02/2019) (cited on page 1).

[3] C. Lewis and J. Müller: *A direct drive wind turbine HTS generator.* 2007 IEEE Power Engineering Society General Meeting, PES:1–8, 2007. DOI: 10.1109/PES.2007.386069 (cited on pages 1, 10).

[4] H. W. Weijers, U. P. Trociewitz, W. D. Markiewicz, J. Jiang, D. Myers, E. E. Hellstrom, A. Xu, J. Jaroszynski, P. Noyes, Y. Viouchkov and D. C. Larbalestier: *High field magnets with HTS conductors.* IEEE Transactions on Applied Superconductivity, **20**(3):576–582, 2010. DOI: 10.1109/TASC.2010.2043080 (cited on pages 1, 10).

[5] S. S. Kalsi: *Applications of high temperature superconductors to electric power equipment.* Wiley-IEEE Press, 2011 (cited on pages 1, 10).

[6] A. M. Luiz, Z. G. Özdemir, A. Çataltepe, Ü. Onbaşlı, G.-J. Lee and Y. Wang: *Applications of High-Tc Superconductivity.* InTechOpen, 2011 (cited on pages 1, 10, 56).

[7] Z. Melhem: *High temperature superconductors (HTS) for energy applications.* Woodhead Publishing Limited, elsevier edition, 2012 (cited on pages 1, 5, 10).

[8] C. Rey: *Superconductors in the Power Grid.* Elsevier Science, 2015 (cited on pages 1, 10).

[9] Magnetic Resonance Imaging. Technical report, Siemens, 2012. DOI: 10.1007/978-3-319-12682-1_18 (cited on page 2).

[10] CERN: Pulling together superconducting magnets. URL: https://home.cern/science/engineering/pulling-together-superconducting-electromagnets (visited on 22/02/2019) (cited on page 2).

[11] M. Landgraf: World's Longest Superconductor Cable. Technical report, AmpaCity, 2012, pages 1–4. URL: http://www.kit.edu/downloads/pi/PI_2012_012_Worlds_Longest_Superconductor_Cable.pdf (cited on page 2).

[12] A. Angrisani Armenio, A. Augieri, G. Celentano, V. Galluzzi, A. Mancini A .and Rufoloni, A. Vannozzi, U. Gambardella, A. Saggese, P. Sessa and S. Pace: *Stability Measurements on YBCO Coated Conductors.* IEEE Transactions on Applied Superconductivity, **18**(2):1293–1296, 2008. DOI: 10.1109/TASC.2008.920835 (cited on pages 2, 16).

[13] S. Honghai and J. Schwartz: *Stability and Quench Behavior of* $YBa_2Cu_3O_{7-x}$ *Coated Conductor at 4.2 K, Self-Field.* IEEE Transactions on Applied Superconductivity, **19**(5):3735–3743, 2009. DOI: 10.1109/TASC.2009.2023674 (cited on pages 2, 16).

[14] F. Trillaud, F. Ayela, A. Devred, M. Fratini, D. Leboeuf and P. Tixador: *Quench propagation ignition using single-mode Diode Laser.* IEEE Transactions on Applied Superconductivity, **15**(2):3648–3651, 2005. DOI: 10.1109/TASC.2005.849381 (cited on pages 3, 16).

[15] J. Pelegrín, E. Martínez, L. A. Angurel, Y. Y. Xie and V. Selvamanickam: *Numerical and experimental analysis of normal zone propagation on 2G HTS wires.* IEEE Transactions on Applied Superconductivity, **21**(3):3041–3044, 2011. DOI: 10.1109/tasc.2010.2084982 (cited on pages 3, 16, 17).

[16] X. Wang, U. P. Trociewitz and J. Schwartz: *Near-adiabatic quench experiments on short* $YBa_2Cu_3O_{7-\delta}$ *coated conductors.* Journal of Applied Physics, **101**(5), 2007. DOI: 10.1063/1.2435804 (cited on pages 3, 16, 17, 80).

[17] C. Lacroix, J. H. Fournier-Lupien, K. McMeekin and F. Sirois: *Normal zone propagation velocity in 2g hts coated conductor with high interfacial resistance.* IEEE Transactions on Applied Superconductivity, **23**(3):3–7, 2013. DOI: 10.1109/TASC.2013.2239696 (cited on pages 3, 16, 17).

[18] C. Lacroix, Y. Lapierre, J. Coulombe and F. Sirois: *High normal zone propagation velocity in second generation high-temperature superconductor coated conductors with a current flow diverter architecture.* Superconductor Science and Technology, **27**(5):055013, 2014. DOI: 10.1088/0953-2048/27/5/055013 (cited on pages 3, 12, 16, 17, 75, 76).

[19] T. P. Orlando and K. A. Delin: *Foundations of Applied Superconductivity.* Addison-Wesley Publishing Company, 1991 (cited on page 5).

[20] M. Cryrot and D. Pavuna: *Introduction to Superconductivity and High-*T_C *Materials.* World Scientific Publishing, Singapore, 2011 (cited on page 5).

[21] L. Dresner: *Stability of Superconductors.* Plenum Press, New York, 1995 (cited on page 5).

[22] Y. Iwasa: *Case studies in superconducting magnets : design and operational issues,* volume 1. Springer, Cambridge, 2. edition edition, 2015. DOI: 10.1007/b112047 (cited on pages 5, 11, 12, 15, 16, 18).

[23] H. K. Onnes: *Further experiments with Liquid Helium G. On the electrical resistance of Pure Metals etc. VI. On the Sudden Change in the Rate at which the Resistance of Mercury Disappears.* Amsterdam Huygens Institute - Royal Netherlands Academy of Arts and Sciences, **14**(II):818–821, 1912 (cited on page 5).

[24] P. J. Ray: *Structural investigation of La(2-x)Sr(x)CuO(4+y) - Following staging as a function of temperature.* Master's thesis, University of Copenhagen, Copenhagen, 2016. DOI: `10.6084/m9.figshare.2075680.v2` (cited on pages 7, 8).

[25] J. Bardeen, L. N. Cooper and J. R. Schrieffer: *Theory of superconductivity.* Physical Review, **108**(5):1175–1204, 1957. DOI: `https://doi.org/10.1103/PhysRev.108.1175` (cited on page 7).

[26] J. Bednorz and K. Muller: *Possible high T_c superconductivity in the Ba-La-Cu-O system.* Zeitschrift für Physik B: Condensed Matter, **64**(2):189–193, 1986. DOI: `10.1007/BF01303701` (cited on page 7).

[27] D. W. Hazelton: *2G HTS Wire Development at SuperPower 2G HTS wire production at SuperPower. The 4th Workshop of the series Accelerator Magnets in High Temperature Superconductors,* Barcelona, 2016. URL: `https://indico.cern.ch/event/588810/contributions/2473740/attachments/1416151/2168254/6_-_DREW_HAZELTON.pdf` (cited on page 9).

[28] SuperPower 2G HTS Wire Specifications. Technical report, SuperPower, 2014. URL: `http://www.superpower-inc.com` (visited on 03/04/2019) (cited on pages 9, 10).

[29] Theva Pro-Line Hts Wire. Technical report, Theva, 2017. URL: `https://www.theva.de/wp-content/uploads/2018/05/171218_THEVA_Broschuere_Pro-Line_generalproperties.pdf` (visited on 21/03/2019) (cited on pages 9, 10).

[30] 2G HTS Tape performance and specifications. Technical report, SuperOx, 2018. URL: `http://www.superox.ru/products/0001.pdf` (visited on 25/02/2019) (cited on pages 9, 10).

[31] Pro-Line: HTS wires produced by e-beam PVD. Technical report, Theva, 2019. URL: `https://www.theva.com` (visited on 03/04/2019) (cited on page 10).

[32] W. D. Markiewicz, D. C. Larbalestier, H. W. Weijers, A. J. Voran, K. W. Pickard, W. R. Sheppard, J. Jaroszynski, A. Xu, R. P. Walsh, J. Lu, A. V. Gavrilin and P. D. Noyes: *Design of a superconducting 32 T magnet with REBCO high field coils.* IEEE Transactions on Applied Superconductivity, **22**(3), 2012. DOI: `10.1109/TASC.2011.2174952` (cited on pages 10, 19).

[33] J. Schwartz, F. Hunte, W. K. Chan, X. F. Gou, X. T. Liu, M. Phillips, Q. V. Le, G. Naderi, M. Turenne and L. Ye: *Status of high temperature superconductor based magnets and the conductors they depend upon.* arXiv preprint:11, 2011. URL: `http://arxiv.org/abs/1108.1634` (cited on page 10).

[34] P. N. Barnes, G. L. Rhoads, J. C. Tolliver, M. D. Sumption and K. W. Schmaeman: *Compact, lightweight, superconducting power generators.* IEEE Transactions on Magnetics, **41**(1 II):268–273, 2005. DOI: `10.1109/TMAG.2004.838984` (cited on page 10).

[35] H. W. Weijers, M. Science, T. Division, N. High and M. Field: *Magnet technology in the NHMFL 32 T HTS-LTS solenoid. The 2nd Workshop an Accelerator Magnet in HTS,* Kyoto. Workshop on HTS Magnet Technology for High Energy Physics, 2014. URL: `https://indico.cern.ch/event/319762/contributions/740342/attachments/616151/847828/Weijers_MagLab_32T_technology_v3.pdf` (cited on pages 10, 19).

[36] H. W. Weijers, W. D. Markiewicz, A. V. Gavrilin, A. J. Voran, Y. L. Viouchkov, S. R. Gundlach, P. D. Noyes, D. V. Abraimov, H. Bai, S. T. Hannahs and T. P. Murphy: *Progress in the Development and Construction of a 32-T Superconducting Magnet.* IEEE Transactions on Applied Superconductivity, **26**(4):1–7, 2016. DOI: `10.1109/TASC.2016.2517022` (cited on page 10).

[37] National High Field Magnet Laboraty: 32 Tesla All-Superconducting Magnet. 2017. URL: `https://nationalmaglab.org/magnet-development/magnet-science-technology/magnet-projects/32-tesla-scm` (visited on 04/04/2019) (cited on pages 10, 15).

[38] S. Hahn, K. Kim, K. Kim, X. Hu, T. Painter, I. Dixon, S. Kim, K. R. Bhattarai, S. Noguchi, J. Jaroszynski and D. C. Larbalestier: *45.5-tesla direct-current magnetic field generated with a high-temperature superconducting magnet.* Nature, **570**(7762):496–499, 2019. DOI: `10.1038/s41586-019-1293-1` (cited on page 10).

[39] C. Lacroix: Discussions on the normal zone proagation velocity. Montréal, 2018 (cited on page 12).

[40] N. Riva: Discussions on the normal zone propagation velocity. Karlsruhe, 2019 (cited on page 12).

[41] J. Rabbers, Jacob: *AC Loss in Superconducting Tapes and Cables.* PhD thesis, University of Twente, 2001 (cited on page 12).

[42] M. N. Wilson: *Superconducting magnets.* Clarendon Press, United Kingdom, 1983 (cited on page 15).

[43] J. Y. Jang, S. Yoon, S. Hahn, Y. J. Hwang, J. Kim, K. H. Shin, K. Cheon, K. Kim, S. In, Y. J. Hong, H. Yeom, H. Lee, S. H. Moon and S. Lee: *Design, construction and 13 K conduction-cooled operation of a 3 T 100 mm stainless steel cladding all-REBCO magnet.* Superconductor Science and Technology, **30**(10), 2017. DOI: 10.1088/1361-6668/aa8354 (cited on page 15).

[44] M. Breschi, L. Trevisani, L. Bottura, A. Devred and F. Trillaud: *Comparing the thermal stability of NbTi and* Nb_3Sn *wires.* Superconductor Science and Technology, **22**(2), 2009. DOI: 10.1088/0953-2048/22/2/025019 (cited on pages 16, 17).

[45] Z. Zhao and Y. Iwasa: *Normal zone propagation in adiabatic superconducting magnets Part 1: Normal zone propagation velocity in superconducting composites.* Cryogenics, **31**(9):817–825, 1991. DOI: 10.1016/0011-2275(91)90141-I (cited on page 16).

[46] T. Verhaege, P. Estop, W. Weber, A. Lacaze, Y. Laumond, P. Bonnet and A. Ansart: *A new class of AC superconducting conductors.* IEEE Transactions on Applied Superconductivity, **3**(1):164–167, 1993. DOI: 10.1109/77.233696 (cited on page 16).

[47] J. H. Fournier-Lupien, C. Lacroix, S. Hellmann, J. Huh, K. Pfeiffer and F. Sirois: *Use of the buffer layers as a current flow diverter in 2G HTS coated conductors.* Superconductor Science and Technology, **31**(12):125019, 2018. DOI: 10.1088/1361-6668/aae2cd (cited on pages 16, 72).

[48] C. Lacroix, F. Sirois and J.-H. F. Lupien: *Engineering of second generation HTS coated conductor architecture to enhance the normal zone propagation velocity in various operating conditions.* Superconductor Science and Technology, **30**(6):064004, 2017. DOI: 10.1088/1361-6668/aa684f (cited on pages 16, 56).

[49] M. Breschi, L. Cavallucci, P. L. Ribani, A. V. Gavrilin and H. W. Weijers: *Analysis of quench in the NHMFL REBCO prototype coils for the 32 T Magnet Project.* Superconductor Science and Technology, **29**(5):055002, 2016. DOI: 10.1088/0953-2048/29/5/055002 (cited on pages 16, 19).

[50] E. A. Young, S. Chappell, I. Falorio and Y. Yang: *Quench characteristics of a Cu-stabilized 2G HTS conductor.* IEEE Transactions on Applied Superconductivity, **21**(3 PART 3):3062–3065, 2011. DOI: 10.1109/TASC.2010.2083615 (cited on page 17).

[51] F. Roy: *Modeling and characterization of coated conductors applied to the design of superconducting fault current limiters.* PhD thesis, École Polytechnique Fédérale de Lausanne, 2010. URL: http://infoscience.epfl.ch/record/148348 (cited on page 17).

[52] J. van Nugteren: *Normal Zone Propagation in a YBCO Superconducting Tape.* Master's thesis, University of Twente, 2012 (cited on pages 17, 109).

[53] Z. Zhong, H. S. Ruiz, L. Lai, Z. Huang, W. Wang and T. Coombs: *Experimental study of the normal zone propagation velocity in double-layer 2G-HTS wires by thermal and electrical methods.* IEEE Transactions on Applied Superconductivity, **25**(3):1–5, 2015. DOI: 10.1109/TASC.2014.2366451 (cited on page 17).

[54] Muons: Fiber Optic Quench Detection via Optimized Rayleigh Scattering in High-field YBCO Accelerator Magnets. Technical report, North Carolina State University, 2015. URL: https://www.osti.gov/servlets/purl/1238212 (cited on page 18).

[55] F. Scurti, S. Ishmael, G. Flanagan and J. Schwartz: *Quench detection for high temperature superconductor magnets: a novel technique based on Rayleigh-backscattering interrogated optical fibers.* Superconductor Science and Technology, **29**(3):03LT01, 2016. DOI: 10.1088/0953-2048/29/3/03LT01 (cited on page 18).

[56] F. Scurti and J. Schwartz: *Optical fiber distributed sensing for high temperature superconductor magnets. 25th International Conference on Optical Fiber Sensors,* volume 10323 of number April 2017, 103238Q. SPIE, 2017. DOI: 10.1117/12.2265947 (cited on page 18).

[57] O. Tsukamoto, J. Maguire, E. Bobrov and Y. Iwasa: *Identification of quench origins in a superconductor with acoustic emission and voltage measurements.* Applied Physics Letters, **39**(2):172–174, 1981. DOI: 10.1063/1.92652 (cited on page 18).

[58] M. Marchevsky, G. Sabbi, H. Bajas and S. Gourlay: *Acoustic emission during quench training of superconducting accelerator magnets.* Cryogenics, **69**:50–57, 2015. DOI: 10.1016/j.cryogenics.2015.03.005 (cited on page 18).

[59] M. Marchevsky and S. A. Gourlay: *Acoustic thermometry for detecting quenches in superconducting coils and conductor stacks.* Applied Physics Letters, **110**(1), 2017. DOI: 10.1063/1.4973466 (cited on page 18).

[60] M. Marchevsky, E. Hershkovitz, X. Wang, S. A. Gourlay and S. Prestemon: *Quench Detection for High-Temperature Superconductor Conductors Using Acoustic Thermometry.* IEEE Transactions on Applied Superconductivity, **28**(4):1–5, 2018. DOI: 10.1109/TASC.2018.2817218 (cited on page 18).

[61] M. Marchevsky, Y.-Y. Xie and V. Selvamanickam: *Quench detection method for 2G HTS wire.* Superconductor Science and Technology, **23**(3):034016, 2010. DOI: 10.1088/0953-2048/23/3/034016 (cited on page 18).

[62] T. Ariyama, T. Takagi, D. Nakayama, E. Sasaki, T. Takao, O. Tsukamoto and T. Matsuoka: *Quench Protection of YBCO Coils: Co-winding Detection Method and Limits to Hot-spot Temperature.* IEEE Transactions on Applied Superconductivity, **8223**(c):1–1, 2016. DOI: 10.1109/TASC.2016.2529838 (cited on page 19).

[63] R. Schmidt, C. Giloux, A. Hilaire, A. Ijspeert and F. Sonnemann: *Protection of LHC superconducting corrector magnets. Proceedings of EPAC*, number 0, pages 2163–2165, Vienna, 2000 (cited on page 19).

[64] F. Rodriguez-Mateos and F. Sonnemann: *Quench heater studics for the LHC magnets. Proceedings of the 2001 Particle Accelerator Conference*, pages 3451–3453, 2002. DOI: 10.1109/pac.2001.988141 (cited on page 19).

[65] K. Dahlerup-Petersen, R. Denz, J. Gomez-Costa, D. Hagedorn, P. Proudlock, F. Rodrinuez-Mateos, R. Schmidt and F. Sonnemann: *The protection system for the superconducting elements of the Large Hadron Collider at CERN. Proceedings of the 1999 Particle Accelerator Conference*, pages 3200–3202, 1999. DOI: 10.1109/pac.1999.792249 (cited on page 19).

[66] P. D. Noyes, W. D. Markiewicz, A. J. Voran, W. R. Sheppard, K. W. Pickard, J. B. Jarvis, H. W. Weijers and A. V. Gavrilin: *Protection heater development for REBCO coils.* IEEE Transactions on Applied Superconductivity, **22**(3):4704204, 2012. DOI: 10.1109/TASC.2012.2188370 (cited on page 19).

[67] E. Ravaioli, V. I. Datskov, C. Giloux, G. Kirby, H. H. Ten Kate and A. P. Verweij: *New, coupling loss induced, quench protection system for superconducting accelerator magnets.* IEEE Transactions on Applied Superconductivity, **24**(3):1–5, 2014. DOI: 10.1109/TASC.2013.2281223 (cited on page 19).

[68] E. Ravaioli, H. Bajas, V. I. Datskov, V. Desbiolles, J. Feuvrier, G. Kirby, M. Maciejewski, G. Sabbi, H. H. Ten Kate and A. P. Verweij: *Protecting a full-scale Nb_3Sn magnet with CLIQ, the new coupling-loss-induced quench system.* IEEE Transactions on Applied Superconductivity, **25**(3), 2015. DOI: 10.1109/TASC.2014.2364892 (cited on page 19).

[69] S. Hahn, D. K. Park, J. Bascuñán and Y. Iwasa: *HTS pancake coils without turn-to-turn insulation.* IEEE Transactions on Applied Superconductivity, **21**(3):
1592–1595, 2011. DOI: 10.1109/TASC.2010.2093492 (cited on page 19).

[70] S. B. Kim, A. Saitou, J. H. Joo and T. Kadota: *The normal-zone propagation properties of the non-insulated HTS coil in cryocooled operation.* Physica C: Superconductivity and its Applications, **471**(21-22):1428–1431, 2011. DOI: 10.1016/j.physc.2011.05.209 (cited on page 20).

[71] S. Hahn, D. K. Park, J. Voccio, J. Bascunan and Y. Iwasa: *No-Insulation (NI) HTS Inserts for >1 GHz LTS/HTS NMR Magnets.* IEEE Transactions on Applied Superconductivity, **22**(3):4302405–4302405, 2012. DOI: 10.1109/TASC.2011.2178976 (cited on page 20).

[72] S. Hahn, Y. Kim, J. Ling, J. Voccio, D. K. Park, J. Bascunan, H. J. Shin, H. Lee and Y. Iwasa: *No-insulation coil under time-varying condition: Magnetic coupling with external coil.* IEEE Transactions on Applied Superconductivity, **23**(3), 2013. DOI: 10.1109/TASC.2013.2240756 (cited on page 20).

[73] Y. Wang, W. K. Chan and J. Schwartz: *Self-protection mechanisms in no-insulation (RE)$ba_2cu_3o_x$ high temperature superconductor pancake coils.* Superconductor Science and Technology, **29**(4):045007, 2016. DOI: 10.1088/0953-2048/29/4/045007 (cited on pages 20, 24, 56, 82).

[74] S. Kim, S. Hahn, K. K. Kim, X. Wang, S. Hahn, Y. Kim, K. K. Kim, K. R. Bhattarai, K. Radcliff, J. Y. Jang, Y. J. Hwang, S. Lee, S. Yoon and S. Hahn: *Quench behavior of a no-insulation coil wound with stainless steel cladding REBCO.* Superconductor Science and Technology, **30**(7):075001, 2017. DOI: 10.1088/1361-6668/aa6a8b (cited on pages 20, 24, 93).

[75] K. L. Kim, Y. H. Choi, D. G. Yang, D. H. Kang, J. H. Kim, H. M. Kim and H. G. Lee: *Analytical and empirical studies on the characteristic resistances of no-insulation GdBCO racetrack pancake coil under various operating currents.* Current Applied Physics, **15**(1):8–13, 2015. DOI: 10.1016/j.cap.2014.10.029 (cited on pages 20, 23, 107, 108).

[76] X. Wang, S. Hahn, Y. Kim, J. Bascuñán, J. Voccio, H. Lee and Y. Iwasa: *Turn-to-turn contact characteristics for an equivalent circuit model of no-insulation ReBCO pancake coil.* Superconductor Science and Technology, **26**(3):035012, 2013. DOI: 10.1088/0953-2048/26/3/035012 (cited on pages 23, 24, 108).

[77] X. Wang, T. Wang, E. Nakada, A. Ishiyama, R. Itoh and S. Noguchi: *Charging behavior in no-insulation REBCO pancake coils.* IEEE Transactions on Applied Superconductivity, **25**(3):1–5, 2015. DOI: 10.1109/TASC.2014.2365623 (cited on page 23).

[78] D. G. Yang, S. Hahn, Y. Kim, K. L. Kim, J. B. Song, J. Bascuñán, H. Lee and Y. Iwasa: *Characteristic resistance of no-insulation and partial-insulation coils with nonuniform current distribution.* IEEE Transactions on Applied Superconductivity, **24**(3), 2014. DOI: 10.1109/TASC.2013.2285103 (cited on pages 23, 24).

[79] H. Jeon, W. S. Lee, J. Kim, G. Baek, S. Jeon, Y. S. Yoon and T. K. Ko: *Investigation of electrical characteristics of no-insulation coil wound with surface-processed HTS tape.* Physica C: Superconductivity and its Applications, **539**:25–29, 2017. DOI: 10.1016/j.physc.2017.06.003 (cited on pages 23, 93).

[80] T. Lecrevisse and Y. Iwasa: *A (RE)BCO Pancake Winding with Metal-as-Insulation.* IEEE Transactions on Applied Superconductivity, **26**(3):1–5, 2016. DOI: 10.1109/TASC.2016.2522638 (cited on page 24).

[81] J. Lu, R. Goddard, K. Han and S. Hahn: *Contact resistance between two REBCO tapes under load and load cycles.* Superconductor Science and Technology, **30**(4):045005, 2017. DOI: 10.1088/1361-6668/aa5b05 (cited on page 24).

[82] J. Lu, J. Levitan, D. McRae and R. Walsh: *Contact resistance between two REBCO tapes: the effects of cyclic loading and surface coating.* Superconductor Science and Technology, **31**(8):085006, 2018. DOI: 10.1088/1361-6668/aacd2d (cited on page 24).

[83] W. K. Chan and J. Schwartz: *Improved stability, magnetic field preservation and recovery speed in (RE)$Ba_2Cu_3O_x$-based no-insulation magnets via a graded-resistance approach.* Superconductor Science and Technology, **30**(7):074007, 2017. DOI: https://doi.org/10.1088/1361-6668/aa6eef (cited on page 24).

[84] D. Liu, H. Yong and Y. Zhou: *Analysis of charging and sudden-discharging characteristics of no-insulation REBCO coil using an electromagnetic coupling model.* AIP Advances, **7**(11):115104, 2017. DOI: 10.1063/1.5001767 (cited on page 24).

[85] D. Liu, W. Zhang and H. Yong: *Thermal stability and mechanical behavior in no-insulation high-temperature superconducting pancake coils.* Superconductor Science and Technology, **31**(8):085010, 2018. DOI: https://doi.org/10.1088/1361-6668/aad00c (cited on page 24).

[86] Y. Wang, H. Song, D. Xu, Z. Y. Li, Z. Jin and Z. Hong: *An equivalent circuit grid model for no-insulation HTS pancake coils.* Superconductor Science and Technology, **28**(4):045017, 2015. DOI: 10.1088/0953-2048/28/4/045017 (cited on pages 24, 82, 102).

[87] J. Kim, S. Yoon, K. Cheon, K. H. Shin, S. Hahn, D. L. Kim, S. G. Lee, H. Lee and S. H. Moon: *Effect of Resistive Metal Cladding of HTS Tape on the Characteristic of No-Insulation Coil.* IEEE Transactions on Applied Superconductivity, **26**(4):1–6, 2016. DOI: 10.1109/TASC.2016.2541687 (cited on page 24).

[88] M. Vollmer and K.-P. Möllmann: *Infrared Thermal Imaging: Fundamentals, Research and Applications.* Wiley, 2010, pages 1–106 (cited on page 27).

[89] W. Demtröder: *Experimentalphysik 2: Elektrizität und Optik (Springer-Lehrbuch) (German Edition).* Springer, 2006 (cited on page 27).

[90] P. Kolodner and J. A. Tyson: *Remote thermal imaging with 0.7-μm spatial resolution using temperature-dependent fluorescent thin flims.* Applied Physics Letters, **42**(1):117–119, 1983. DOI: 10.1063/1.93766 (cited on pages 27, 30, 32).

[91] P. Kolodner and J. A. Tyson: *Microscopic fluorescent imaging of surface temperature profiles with 0.01 ° C resolution.* Applied Physics Letters, **40**(9):782–784, 1982. DOI: 10.1063/1.93258 (cited on pages 27, 30, 32).

[92] O. Haugen, T. H. Johansen, H. Chen, V. Yurchenko, P. Vase, D. Winkler, B. A. Davidson, G. Testa, E. Sarnelli and E. Altshuler: *High resolution thermal imaging of hotspots in superconducting films.* IEEE Transactions on Applied Superconductivity, **17**(2):3215–3218, 2007. DOI: 10.1109/TASC.2007.899341 (cited on pages 28, 30, 32).

[93] Ø. Haugen and T. Johansen: *Temperature dependent photoluminescence down to 4.2K in EuTFC.* Journal of Luminescence, **128**(9):1479–1483, 2008. DOI: 10.1016/j.jlumin.2008.02.007 (cited on pages 28, 30, 32, 35, 36).

[94] K. Nara, N. Yamada and H. Kato: *Assessment of Commecial Fluorescent Paints on Their Applicability to the Cryogenic Thermography System. AIP Conference Proceedings*, volume 823, pages 441–448. AIP, 2006. DOI: 10.1063/1.2202446 (cited on page 28).

[95] A. Ishiyama, M. Tsuchiya, H. Ueda and Y. Shiohara: *Assessment of cryogenic thermography system using commercial fluorescent paints on their applicability to visualization of normal-zone propagation in YBCO coated conductor.* IEEE Transactions on Applied Superconductivity, **17**(2):3765–3768, 2007. DOI: 10.1109/TASC.2007.899124 (cited on page 28).

[96] *Basic principles of fluorescence spectroscopy. Handbook of Fluorescence Spectroscopy and Imaging.* John Wiley & Sons, Ltd, 2011. Chapter 1, pages 1–30. DOI: 10.1002/9783527633500.ch1 (cited on page 30).

[97] J. R. Lakowicz: *Introduction to Fluorescence Spectroscopy*, volume 689 of number 3. PerkinElmer, 2011, pages 93–136. DOI: 10.1007/978-1-60761-950-5_7 (cited on page 30).

[98] *The Molecular Probes Handbook—A Guide to Fluorescent Probes and Labeling Technologies. Thermo Fischer.* 11th edition. Invitrogen by Thermo Fischer, 11th edition, 2010. Chapter Fluorescence Fundamentals, pages 3–8 (cited on page 30).

[99] N. Baroni: Fluorescent spectroscopy measurements. 2017 (cited on pages 32, 37).

[100] G. Hampel, P. Kolodner, P. L. Gammel, P. A. Polakos, E. de Obaldia, P. M. Mankiewich, A. Anderson, R. Slattery, D. Zhang, G. C. Liang and C. F. Shih: *High power failure of superconducting microwave filters: investigation by means of thermal imaging.* Applied Physics Letters, **69**(4):571–573, 1996. DOI: 10.1063/1.117790 (cited on page 32).

[101] LED Engin: *365nm UV LED Gen 2 Emitter LZ1-00UV00.* OSRAM. 2016 (cited on pages 33, 55).

[102] L. D. Barton: Fluorescent Microthermographic Imaging, (1993). URL: https://www.osti.gov/biblio/10186283 (cited on pages 35, 39).

[103] R. Popov: Atomic force microscopy measurements. 2016 (cited on page 37).

[104] M. Jakoby: Temperature dependent fluorescence spectroscopy measurements. 2017 (cited on pages 37, 39).

[105] A. Eicher: *Optische und elektrische Untersuchung des Quench- und Recoveryverhaltens von Dünnschichtsupraleitern.* Master's thesis, Karlsruhe Institute of Technology, 2015 (cited on page 50).

[106] S. Hellmann: *Development on Superconducting Current Limiting Transformers.* PhD thesis, Karlsruhe Institute of Technology, Karlsruhe, 2018. DOI: 10.5445/IR/1000088555 (cited on pages 50, 132).

[107] Horizon2020: Fastgrid Project. 2018. URL: https://www.fastgrid-h2020.eu/ (visited on 22/06/2018) (cited on pages 52, 74).

[108] M. K. Wu, J. R. Ashburn, C. J. Torng, P. H. Hor, R. L. Meng, L. Gao, Z. J. Huang, Y. Q. Wang and C. W. Chu: *Superconductivity at 93 K in a new mixed-phase Y-Ba-Cu-O compound system at ambient pressure.* Physical Review Letters, **58**(9):908–910, 1987. DOI: 10.1103/PhysRevLett.58.908 (cited on page 56).

[109] D. Colangelo and B. Dutoit: *Impact of the normal zone propagation velocity of high-temperature superconducting coated conductors on resistive fault current limiters.* IEEE Transactions on Applied Superconductivity, **25**(2), 2015. DOI: 10.1109/TASC.2015.2396935 (cited on page 56).

[110] C. Lacroix and F. Sirois: *Concept of a current flow diverter for accelerating the normal zone propagation velocity in 2G HTS coated conductors.* Superconductor Science and Technology, **27**(12):129501, 2014. DOI: 10.1088/0953-2048/27/12/129501 (cited on pages 61, 72).

[111] X. Granados, R. Guzmán, E. Bartolomé, A. Palau, A. Calleja, A. Usoskin, J. Arbiol, X. Obradors, V. R. Vlad, T. Puig and M. Aklalouch: *Magnetic and structural characterization of inkjet-printed* $^{TFA}YBa_2Cu_3O_{7-x}$ $/^{MOD}$ $CZO/^{ABAD}$ *YSZ/SS coated conductors.* Superconductor Science and Technology, **26**(12):125004, 2013. DOI: 10.1088/0953-2048/26/12/125004 (cited on pages 61, 73).

[112] V. R. Vlad, E. Bartolome, M. Vilardell, A. Calleja, A. Meledin, X. Obradors, T. Puig, S. Ricart, G. Van Tendeloo, A. Usoskin, S. Lee, V. Petrykin and A. Molodyk: *Inkjet Printing Multideposited YBCO on CGO/LMO/MgO/Y_2O_3/ Al_2O_3/Hastelloy Tape for 2G-Coated Conductors.* IEEE Transactions on Applied Superconductivity, **28**(4):1–5, 2018. DOI: 10.1109/tasc.2018.2808403 (cited on pages 61, 73).

[113] A. Calleja, M. Vilardell, V. R. Vlad, X. Obradors, T. Puig Molina, S. Ricart and X. Granados: Método de fabricación de láminas delgadas de óxidos epitaxiales mediante impresión de chorro de tinta. Patent (2 601 487). 2017. URL: http://hdl.handle.net/10261/175465 (cited on page 73).

[114] P. Barusco: Scanning Hall Probe Microscopy. Barcelona, 2018 (cited on page 73).

[115] N. T. Nguyen, C Barnier and P Tixador: *Optical investigation of the quenching of coated conductors.* Journal of Physics: Conference Series, **234**(3):032058, 2010. DOI: 10.1088/1742-6596/234/3/032058 (cited on page 80).

[116] W. K. Chan and J. Schwartz: *A Hierarchical Three-Dimensional Multiscale Electro–Magneto–Thermal Model of Quenching in $REBa_2Cu_3O_{7-\delta}$ Coated-Conductor-Based Coils.* IEEE Transactions on Applied Superconductivity, **22**(5):4706010–4706010, 2012. DOI: 10.1109/TASC.2012.2198647 (cited on page 82).

[117] T. Wang, S. Noguchi, X. Wang, I. Arakawa, K. Minami, K. Monma, A. Ishiyama, S. Hahn and Y. Iwasa: *Analyses of transient behaviors of no-insulation REBCO pancake coils during sudden discharging and overcurrent.* IEEE Transactions on Applied Superconductivity, **25**(3):1–9, 2015. DOI: 10.1109/TASC.2015.2393058 (cited on pages 82, 102).

[118] Y. Wang and H. Song: *Influence of turn-to-turn resistivity and coil geometrical size on charging characteristics of no-electrical-insulation REBCO pancake coils.* Superconductor Science and Technology, **29**(7):075006, 2016. DOI: 10.1088/0953-2048/29/7/075006 (cited on page 82).

[119] T. Oki, A. Ikeda, TaoWang, A. Ishiyama, S. Noguchi, K. Monma, T. Watanabe and S. Nagaya: *Evaluation on Quench Protection for No-insulation REBCO Pancake Coil.* IEEE Transactions on Applied Superconductivity, **26**(4), 2016. DOI: 10.1109/TASC.2016.2540001 (cited on page 82).

[120] Y. Suetomi, K. Yanagisawa, H. Nakagome, M. Hamada, H. Maeda and Y. Yanagisawa: *Mechanism of notable difference in the field delay times of no-insulation layer-wound and pancake-wound REBCO coils.* Superconductor Science and Technology, **29**(10), 2016. DOI: 10.1088/0953-2048/29/10/105002 (cited on page 82).

[121] Y. Liu, J. Ou, F. Grilli, F. Schreiner, V. M. R. Zermeno, M. Zhang and M. Noe: *Comparison of 2D simulation models to estimate the critical current of a coated superconducting coil.* Superconductor Science and Technology, **32**(1):014001, 2018. DOI: 10.1088/1361-6668/aae960 (cited on page 82).

[122] M. Lao: Angular dependence of J_C measurement. Technical report, Karlsruhe Institute of Technology, Kalrsurhe, 2018 (cited on pages 83, 132, 133).

[123] M. Lao, J. Hänisch, S. Kauffmann-Weiss, R. Gehring, H. Fillinger, A. Drechsler and B. Holzapfel: *High current variable temperature electrical characterization system for superconducting wires and tapes with continuous sample rotation in a split coil magnet.* Review of Scientific Instruments, **90**(1):015106, 2019. DOI: 10.1063/1.5078447 (cited on page 83).

[124] S. Otten, A. Kario, A. Kling and W. Goldacker: *Bending properties of different REBCO coated conductor tapes and Roebel cables at $T = 77\ K$.* Superconductor Science and Technology, **29**(12):125003, 2016. DOI: 10.1088/0953-2048/29/12/125003 (cited on page 89).

[125] K. Altmann: Microscopy images of HTS tapes. 2018 (cited on page 91).

[126] H. C. Jo, S. Choi, J. B. Na, J. Y. Jang, Y. J. Hwang, H. J. Kim, M. C. Ahn, Y. D. Chung, H. M. Kim, Y. S. Yoon, K. S. Ryu, Y. C. Kim, H. Lee and T. K. Ko: *Characteristic comparison for the various winding methods of HTS magnets.* IEEE Transactions on Applied Superconductivity, **22**(3), 2012. DOI: 10.1109/TASC.2012.2186549 (cited on page 93).

[127] K. Yanagisawa, S. Iguchi, Y. Xu, J. Li, A. T. Saito, H. Nakagome, T. Takao, S. Matsumoto, M. Hamada and Y. Yanagisawa: *A Long Charging Delay for a No-Insulation REBCO Layer-Wound Coil and Its Influence on Operation With Outer LTS Coils.* IEEE Transactions on Applied Superconductivity, **26**(4):2–5, 2016. DOI: 10.1109/TASC.2016.2515540 (cited on page 93).

[128] S. Yazaki, A. Karasawa, T. Kotoyori, A. Ishiyama and N. Miyahara: *Critical Current Degradation in High-Temperature Superconducting Tapes Caused by Temperature Rise.* Applied Superconductivity, IEEE Transactions, **23**(3):4602304, 2013. DOI: 10.1109/TASC.2013.2244157 (cited on page 93).

[129] R. Gyuráki, F. Sirois and F. Grilli: *High-speed fluorescent thermal imaging of quench propagation in high temperature superconductor tapes.* Superconductor Science and Technology, **31**(3):34003, 2018. DOI: 10.1088/1361-6668/aaa703 (cited on pages 96, 97).

[130] S. J. Otten: *Characterisation of REBCO Roebel cables.* PhD thesis, Karlsruhe Institute of Technology, 2019 (cited on page 99).

[131] N. J. Simon, E. S. Drexler and R. P. Reed: *Properties of copper and copper alloys at cryogenic temperature.* Volume NIST monograph 177. National Bureau of Standards, 1992. Chapter 7, pages 1–15 (cited on page 109).

Karlsruher Schriftenreihe zur Supraleitung

Karlsruher Institut für Technologie (KIT) | ISSN 1869-1765

Band 001 **Christian Schacherer**
Theoretische und experimentelle Untersuchungen zur Entwicklung supraleitender resistiver Strombegrenzer.
ISBN 978-3-86644-412-6

Band 002 **Alexander Winkler**
Transient behaviour of ITER poloidal field coils.
ISBN 978-3-86644-595-6

Band 003 **André Berger**
Entwicklung supraleitender, strombegrenzender Transformatoren.
ISBN 978-3-86644-637-3

Band 004 **Christoph Kaiser**
High quality Nb/Al-AlOx/Nb Josephson junctions. Technological development and macroscopic quantum experiments.
ISBN 978-3-86644-651-9

Band 005 **Gerd Hammer**
Untersuchung der Eigenschaften von planaren Mikrowellen-resonatoren für Kinetic-Inductance Detektoren bei 4,2 K.
ISBN 978-3-86644-715-8

Band 006 **Olaf Mäder**
Simulationen und Experimente zum Stabilitätsverhalten von HTSL-Bandleitern.
ISBN 978-3-86644-868-1

Band 007 **Christian Barth**
High Temperature Superconductor Cable Concepts for Fusion Magnets.
ISBN 978-3-7315-0065-0

Die Bände sind unter www.ksp.kit.edu als PDF frei verfügbar oder als Druckausgabe bestellbar.

Band 008 **Axel Stockhausen**
Optimization of Hot-Electron Bolometers for THz Radiation.
ISBN 978-3-7315-0066-7

Band 009 **Petra Thoma**
Ultra-fast $YBa_2Cu_3O_{7-x}$ direct detectors for the THz frequency range.
ISBN 978-3-7315-0070-4

Band 010 **Dagmar Henrich**
Influence of Material and Geometry on the Performance of Superconducting Nanowire Single-Photon Detectors.
ISBN 978-3-7315-0092-6

Band 011 **Alexander Scheuring**
Ultrabreitbandige Strahlungseinkopplung in THz-Detektoren.
ISBN 978-3-7315-0102-2

Band 012 **Markus Rösch**
Development of lumped element kinetic inductance detectors for mm-wave astronomy at the IRAM 30 m telescope.
ISBN 978-3-7315-0110-7

Band 013 **Johannes Maximilian Meckbach**
Superconducting Multilayer Technology for Josephson Devices.
ISBN 978-3-7315-0122-0

Band 014 **Enrico Rizzo**
Simulations for the optimization of High Temperature Superconductor current leads for nuclear fusion applications.
ISBN 978-3-7315-0132-9

Band 015 **Philipp Krüger**
Optimisation of hysteretic losses in high-temperature superconducting wires.
ISBN 978-3-7315-0185-5

Band 016 **Matthias Hofherr**
Real-time imaging systems for superconducting nanowire single-photon detector arrays.
ISBN 978-3-7315-0229-6

Band 017 **Oliver Näckel**
Development of an Air Coil Superconducting Fault Current Limiter.
ISBN 978-3-7315-0526-6

Band 018 **Christoph M. Bayer**
Characterization of High Temperature Superconductor Cables for Magnet Toroidal Field Coils of the DEMO Fusion Power Plant.
ISBN 978-3-7315-0605-8

Band 019 **Shengnan Zou**
Magnetization of High Temperature Superconducting Trapped-Field Magnets.
ISBN 978-3-7315-0715-4

Band 020 **Ilya Charaev**
Improving the Spectral Bandwidth of Superconducting Nanowire Single-Photon Detectors (SNSPDs).
ISBN 978-3-7315-0745-1

Band 021 **Juliane Raasch**
Electrical-field sensitive $YBa_2Cu_3O_{7-x}$ detectors for real-time monitoring of picosecond THz pulses.
ISBN 978-3-7315-0786-4

Band 022 **Yingzhen Liu**
Design of a superconducting DC wind generator.
ISBN 978-3-7315-0796-3

Band 023 **Sebastian Hellmann**
Research and Technology Development on Superconducting Current Limiting Transformers.
ISBN 978-3-7315-0804-5

Die Bände sind unter www.ksp.kit.edu als PDF frei verfügbar oder als Druckausgabe bestellbar.

Band 024 **Simon J. Otten**
Characterisation of REBCO Roebel cables.
ISBN 978-3-7315-0904-2

Band 025 **Julia Brandel**
Supraleitende Einzelphotonenzähler: Optimierung der Zeitauflösung und Anwendungsbeispiele aus der Spektroskopie.
ISBN 978-3-7315-0917-2

Band 026 **Dustin Kottonau, Eugen Shabagin, Wescley T. B. de Sousa, Jörn Geisbüsch, Mathias Noe, Hanno Stagge, Simon Fechner, Hannes Woiton, Thomas Küsters**
Bewertung des Einsatzes supraleitender 380-kV-Kabel.
ISBN 978-3-7315-0927-1

Band 027 **Steffen Dörner**
Multifrequenzausleseverfahren von supraleitenden Einzelphotonen-Detektoren.
ISBN 978-3-7315-0961-5

Band 028 **Michael Merker**
Superconducting integrated THz receiver.
ISBN 978-3-7315-0970-7

Band 029 **Wolfgang-Gustav Ekkehart Schmidt**
Superconducting Nanowire Single-Photon Detectors for Quantum Photonic Integrated Circuits on GaAs.
ISBN 978-3-7315-0980-6

Band 030 **Dustin Kottonau, Eugen Shabagin, Wescley de Sousa, Jörn Geisbüsch, Mathias Noe, Hanno Stagge, Simon Fechner, Hannes Woiton, Thomas Küsters**
Evaluation of the Use of Superconducting 380 kV Cable.
ISBN 978-3-7315-1026-0

Band 031 **Alan Preuß**
Development of high-temperature superconductor cables for high direct current applications.
ISBN 978-3-7315-1041-3

Band 032 **Roland Gyuráki**
Fluorescent thermal imaging method for investigating transient effects in high-temperature superconductor tapes and coils.
ISBN 978-3-7315-1064-2